MEDIA AND COMMUNICATIONS - TECHNOLOGIES,
POLICIES AND CHALLENGES

MOBILE PHONES

TECHNOLOGY, NETWORKS AND USER ISSUES

MEDIA AND COMMUNICATIONS - TECHNOLOGIES, POLICIES AND CHALLENGES

Additional books in this series can be found on Nova's website under the Series tab.

Additional E-books in this series can be found on Nova's website under the E-books tab.

MEDIA AND COMMUNICATIONS - TECHNOLOGIES,
POLICIES AND CHALLENGES

MOBILE PHONES

TECHNOLOGY, NETWORKS AND USER ISSUES

MICAELA C. BARNES
AND
NEIL P. MEYERS
EDITORS

Nova Science Publishers, Inc.
New York

Copyright © 2011 by Nova Science Publishers, Inc.

All rights reserved. No part of this book may be reproduced, stored in a retrieval system or transmitted in any form or by any means: electronic, electrostatic, magnetic, tape, mechanical photocopying, recording or otherwise without the written permission of the Publisher.

For permission to use material from this book please contact us:
Telephone 631-231-7269; Fax 631-231-8175
Web Site: http://www.novapublishers.com

NOTICE TO THE READER

The Publisher has taken reasonable care in the preparation of this book, but makes no expressed or implied warranty of any kind and assumes no responsibility for any errors or omissions. No liability is assumed for incidental or consequential damages in connection with or arising out of information contained in this book. The Publisher shall not be liable for any special, consequential, or exemplary damages resulting, in whole or in part, from the readers' use of, or reliance upon, this material. Any parts of this book based on government reports are so indicated and copyright is claimed for those parts to the extent applicable to compilations of such works.

Independent verification should be sought for any data, advice or recommendations contained in this book. In addition, no responsibility is assumed by the publisher for any injury and/or damage to persons or property arising from any methods, products, instructions, ideas or otherwise contained in this publication.

This publication is designed to provide accurate and authoritative information with regard to the subject matter covered herein. It is sold with the clear understanding that the Publisher is not engaged in rendering legal or any other professional services. If legal or any other expert assistance is required, the services of a competent person should be sought. FROM A DECLARATION OF PARTICIPANTS JOINTLY ADOPTED BY A COMMITTEE OF THE AMERICAN BAR ASSOCIATION AND A COMMITTEE OF PUBLISHERS.

Additional color graphics may be available in the e-book version of this book.

Library of Congress Cataloging-in-Publication Data

Mobile phones : technology, networks, and user issues / editors, Micaela C. Barnes and Neil P. Meyers.
    p. cm.
  Includes index.
  ISBN 978-1-61209-247-8 (hardcover)
 1. Cell phone systems. 2. Cell phone systems--Health aspects. I. Barnes, Micaela C. II. Meyers, Neil P.
  TK5103.2.M595 2010
  384.5'34--dc22
                            2010047055

Published by Nova Science Publishers, Inc. † New York

CONTENTS

Preface		vii
Chapter 1	Biological Impacts, Action Mechanisms, Dosimetry and Protection Issues of Mobile Telephony Radiation *Dimitris J. Panagopoulos*	1
Chapter 2	Mobile Phones, Multimedia and Communicability: Design, Technology Evolution, Networks and User Issues *Francisco V. Cipolla-Ficarra*	55
Chapter 3	Increased Genetic Damage due to Mobile Telephone Radiations *A. S. Yadav, Manoj Kumar Sharma and Shweta Yadav*	95
Chapter 4	Designing Mobile Phone Interfaces for Collaborative Learning in Everyday Life *Júlio Cesar dos Reis, Rodrigo Bonacin and Maria Cecília Martins*	133
Chapter 5	International Expansion of European Operators: A Descriptive Study *Lucio Fuentelsaz, Elisabet Garrido and Juan Pablo Maicas*	155
Chapter 6	M-Healthcare: Combining Healthcare, Health Management, and the Social Support of the Virtual Community *Wen-Yuan Jen*	183
Chapter 7	OLS - Opportunistic Localization System for Smart Phones Devices *Maarten Weyn and Martin Klepal*	191
Chapter 8	Mobile Phones and Inappropriate Content *J. G. Phillips, P. Ostojic and A. Blaszczynski*	227
Index		243

PREFACE

The fast evolution of the technology, social network services and mobile platforms have transformed the traditional notions of community and intercultural communication. This new book presents topical research in the study of the technology, network and user issues in mobile phones today. Topics discussed include increased genetic damage due to mobile telephone radiation; mobile phone interfaces for collaborative learning in everyday life; the internationalization process of European operators and mobile health services improving healthcare through mobile technology. (Imprint: Nova Press).

Chapter 1 – Different kinds of biological effects of mobile telephony radiation have been already confirmed by different experimenters, while a lot of contradictory results are also reported. In spite of any uncertainty, some of the recent results reporting effects show a distinct agreement between them, although with different biological models and under different laboratory conditions. Such results of exceptional importance and mutual similarity are those reporting DNA damage or oxidative stress induction on reproductive cells of different organisms, resulting in decreased fertility and reproduction. This distinct similarity among results of different experimental studies makes unlikely the possibility that these results could be wrong. This chapter analyzes and resumes the authors' experimental findings of DNA damage on insect reproductive cells by Global System for Mobile telecommunications (GSM) radiations, compares them with similar recent results on mammalian - human infertility and discusses the possible connection between these findings and other reports regarding tumour induction, symptoms of unwellness, or declinations of bird and insect populations. A possible biochemical explanation of the reported effects at the cellular level is attempted. Since microwave radiations are non-ionizing and therefore unable to break chemical bonds, indirect ways of DNA damage are discussed, through enhancement of free radical and reactive oxygen species (ROS) formation, or irregular release of hydrolytic enzymes. Such events can be initiated by alterations of intracellular ionic concentrations after irregular gating of electrosensitive channels on the cell membranes according to the Ion Forced-Vibration mechanism that the authors have previously proposed. This biophysical mechanism seems to be realistic, since it is able to explain all of the reported biological effects associated with electromagnetic fields (EMFs) exposure, including the so-called "windows" of increased bioactivity reported since many years but remained unexplained so far, and recorded also in the authors' recent experiments in regards to GSM radiation exposure. The chapter discusses also, an important dosimetry issue, regarding the use of Specific Absorption Rate (SAR), a quantity introduced to describe temperature increases within biological tissue (thermal effects), while the vast majority of the recorded biological effects are non-thermal. Finally the chapter attempts to propose some basic precautions and a different way of mobile telephony base station antennas network design, in order to minimize the exposure of human population and reduce significantly the current exposure limits in order to account for the reported non thermal biological effects.

Chapter 2 – In the authors' research work the authors focus on the users who have a greater difficulty in understanding the functionality of mobile phones with the triad of communicability, usability and usefulness. It is a detailed study which starts with the models of the 90s and the first decade of the new millennium, and tries to underline the area where the formal and factual sciences interact. To this end an unprecedented evaluation methodology is established where there is a technological and communicative intersection. The participation of real users allows us to put into use a series of heuristic research techniques to face these problems from several points of view of interactive and communicational design. In this work the authors make a special analysis of the subject of the choice of the random samples inside descriptive statistics, since the authors are working with it to cut down costs in the methods and evaluation of the quality in the interactive systems, where the mobile multimedia phones are included, for example. Additionally, the obtained results allow not only to know the weak and strong points of the current multimedia mobile phones but the authors set up a little vademecum where the sociological and technological aspects of the users in the coming years are summed up.

Chapter 3 – Mobile telephony is becoming increasingly integrated into everyday life. Recently mobile communication has experienced a vast expansion all over the world. With the advent of newer and newer technologies, particularly in the telecommunication, has made humans highly vulnerable to both ionizing and non-ionizing electromagnetic radiations (EMR) along with the natural background radiations. Electromagnetic fields now blanket the earth with a huge network of systems which emit these radiations in to the environment. The health risk issues related to the application of telecommunication systems became a worldwide concern. International Agency for Research on Cancer (IARC) has classified low frequency electromagnetic field as a possible carcinogen. New concept of how radiations interact with biological system is bound to suggest new approaches and new avenues for development of novel radio sensitizers for tumour treatment. Exposure of living cells to EMR activates genetic cascade of signaling events leading to cellular damage, primarily through a spectrum of lesions induced in DNA. This damage to living cells by radiations takes place at molecular level and can induce genetic instability. A number of biological effects induced by man-made Electromagnetic fields and radiations of different frequencies, including digital mobile telephony have already been documented by many research groups. During the present investigation 250 individuals, irrespective of age/sex/caste have been taken. Subjects exposed to mobile phone radiations (110) and healthy controls (140) matched with respect to age, sex and socio-economic status comprised the material. During the study no incidence of cancer or other diseases in persons exposed to EMR through mobile telephones was found. Comet Assay and Micronucleus (MN) Assay were used as the biomarkers of choice for evaluating genotoxicity. The mean frequency of micronucleated cells (MNC), total micronuclei (TMN) and binucleated cells (BN) showed an increase and the difference was found to be statistically significant with respect to controls. Similarly, statistically significant difference was observed in the entire comet parameters between exposed and control subjects, thereby indicating DNA damage in the form of comet length, tail length, percent DNA in tail and Olive tail movement. The subjects exposed to mobile phone radiations showed a higher degree of genetic damage in DNA as compared to the control group. A positive correlation was observed between the duration of exposure and the extent of genetic damage in the mobile phone user group.

Chapter 4 – Literature points out new perspectives for collaborative learning through the use of mobile devices. This subject has being widely discussed by education and computer science researchers, showing as one of the strands for the future of (in)formal education. By using mobile devices, learning could be developed through collaborative interactions at any time or location. This represents a new possibility for people "learning while doing" their everyday activities. Moreover, the advent of mobile phones has created new opportunities that go beyond simple communication acts; their software interfaces have a primary role in enabling the collaboration among the evolved parties. The Authors propose a novel approach which uses mobile collaborative learning for supporting everyday life tasks in general. In order to enrich the mobile collaborative learning, software applications should better explore the interfaces and multimedia resources available in current mobile devices. The mobile interfaces could support and boost situations that lead to learning, however it is essential to minimize the interaction difficulties, and maximize the learning activities itself. To achieve that, this work presents a design proposal and prototype of mobile phone interfaces for supporting mobile collaborative discussion, illustrating the ideas and the design decisions. This new approach aims to the enrichment of mobile interfaces, employing different medias and forms of interaction for the purpose of to constitute "wireless" communities of knowledge sharing about any issue or topic; thus stimulating and promoting the constitution of "communities of practice" through interaction, in which members can share common problems and/or work domain. In the proposed design, the resources of the interfaces are essentials to enable users to explain better their ideas; for that the paper presents multimedia interfaces to share images, sounds, and videos during the discussions. The authors also present a discussion about the impacts of this approach for informal education, and preliminary results from a qualitative analysis with real users.

Chapter 5 – This chapter describes the internationalization process followed by the main European mobile operators from 1998 to 2008. The authors measure the degree of internationalization by the number of countries in which they operate in European OECD countries and the rest of the world. In this chapter, the authors relate the international diversification to (i) the history of wireless communications in Europe, (ii) the evolution in the mode of market entry and, finally, (iii) the rebranding process followed by international wireless groups. This chapter aims to serve as a guide to the different patterns of internationalization of the main European operators.

Chapter 6 – Mobile health services (m-health services) offered through mobile technology have contributed greatly to improved healthcare. At present, most m-health services focus on patient care and treatment; unfortunately, the potential benefit of mobile technology to preventive medicine has not yet been adequately explored. As the population ages and instances of obesity and obesity-related illness among the general population increase, preventative medicine services delivered by m-health systems may help relieve pressure on limited healthcare resources. M-health services are available 24/7 and, more importantly for preventative medicine, have the potential to create virtual support groups that will improve preventative medicine effectiveness. This chapter describes the potential of m-health technology to deliver preventative medicine information and assistance and to create effective support groups among individuals with common health concerns.

Chapter 7 – People are eager to locate their peers and stay connected with them. OLS makes that simple. OLS's core technology is an opportunistic location system based on smart phone devices. Its main advantage is that it seamlessly works throughout heterogeneous environments including indoors as opposed to GPS based systems being available only outdoors under the unobstructed sky. OLS is a phone-centric localization system which grasps at any location related information readily available in the mobile phone. In contrast to most of the competing indoor localization systems, OLS does not require a fixed dedicated infrastructure to be installed in the environment making OLS a truly ubiquitous localization service. The latest version of OLS strives to reduce the system ownership cost by adopting a patent covered self-calibration mechanism minimizing the system installation and maintenance cost even further. OLS's architecture migrated from the original client-server to the current service oriented architecture to cope with increasing demands on reusability across various environments and platforms and to scale up to service a large number of various clients. The location related information used for the estimation of mobile device location are existing signals in the environment, which can be sensed by a smart phone. The readily available information, depending on the phone capability, is typically a subset or all of the following: the GSM/UMTS signal strength, WiFi signal strength, GPS, reading from embedded accelerometers and Bluetooth proximity information. The reliability and availability of input information depends strongly on the actual character of the mobile client physical environment. When the client is outdoors under the unobstructed sky, GPS is a favorable choice typically combined with pedometer data derived from the 3D acceleration and compass measurement. When the client is in a dense urban and indoor environment the GSM/UMTS and WiFi signal strength combined with pedometer data usually performs best. If indoor floor plan layouts are available, the map filtering algorithm can further contribute to the location estimation. However, the information about the actual type of environment is not available and proper importance weighting of all input information in the fusion engine is paramount. The chapter will describe the developed OLS and particularly give insight of the OLS core technology that is the adaptive fusion of location related information readily available in smart phones. The experiments carried out, shows when considering all location related information for the object location estimation namely the GSM and WiFi signal strength, GPS, PDR and map filtering, the presented opportunistic localization system achieves a mean error of 2.73 m and a correct floor detection of 93% in common environments. The achieved accuracy and robustness of the system should be sufficient for most of location aware services and application, therefore having a potential of enabling truly ubiquitous location aware computing.

Chapter 8 – The mobile phone is touted to bridge the digital divide, with mobile phones being possessed by most individuals regardless of their age or ability. Improvements in phone and network capability have meant that most web content is now accessible anywhere by anyone at anytime. The degree to which the digital divide is being bridged has caused problems for legislators seeking to control access to specific content or activities (e.g. gambling; pornography) by sections of the community, particularly those most vulnerable (i.e. children and adolescents). The process of access to inappropriate content, or the excessive participation in activities, actually requires the transmission of a number of messages. This chapter discusses methods of controlling access to inappropriate content and restricting excessive activities. Providers of inappropriate content can be blocked at the source. Regulators have had success prosecuting providers of specific content (e.g. missed call scams), but jamming systems lack specificity. The use of contracts has tended to limit access to users of legal age, but may be circumvented by the use of debit cards, disposable and prepaid phones. Biometric systems have been under consideration to determine age, but are not foolproof. Attempts to control inappropriate content are compromised by providers operating off-shore.

In: Mobile Phones: Technology, Networks and User Issues
Editors: Micaela C. Barnes et al., pp. 1-54
ISBN: 978-61209-247-8
©2011 Nova Science Publishers, Inc.

Chapter 1

BIOLOGICAL IMPACTS, ACTION MECHANISMS, DOSIMETRY AND PROTECTION ISSUES OF MOBILE TELEPHONY RADIATION

Dimitris J. Panagopoulos[*]

University of Athens, Faculty of Biology,
Department of Cell Biology and Biophysics, Athens, Greece
Radiation and Environmental Biophysics Research Centre, Athens, Greece

ABSTRACT

Different kinds of biological effects of mobile telephony radiation have been already confirmed by different experimenters, while a lot of contradictory results are also reported. In spite of any uncertainty, some of the recent results reporting effects show a distinct agreement between them, although with different biological models and under different laboratory conditions. Such results of exceptional importance and mutual similarity are those reporting DNA damage or oxidative stress induction on reproductive cells of different organisms, resulting in decreased fertility and reproduction. This distinct similarity among results of different experimental studies makes unlikely the possibility that these results could be wrong. This chapter analyzes and resumes our experimental findings of DNA damage on insect reproductive cells by Global System for Mobile telecommunications (GSM) radiations, compares them with similar recent results on mammalian - human infertility and discusses the possible connection between these findings and other reports regarding tumour induction, symptoms of unwellness, or declinations of bird and insect populations. A possible biochemical explanation of the reported effects at the cellular level is attempted. Since microwave radiations are non-ionizing and therefore unable to break chemical bonds, indirect ways of DNA damage are discussed, through enhancement of free radical and reactive oxygen species (ROS) formation, or irregular release of hydrolytic enzymes. Such events can be initiated by alterations of intracellular ionic concentrations after irregular gating of electrosensitive

---

[*] Correspondence: 1) Dr. Dimitris J. Panagopoulos, Department of Cell Biology and Biophysics, Faculty of Biology, University of Athens, Panepistimiopolis, 15784, Athens, Greece, Fax: +30210 7274742, Phone: +30210 7274273. E-mail: dpanagop@biol.uoa.gr, 2) Dr. Dimitris J. Panagopoulos, Radiation and Environmental Biophysics Research Centre, 79 Ch. Trikoupi str., 10681 Athens, Greece., E-mail: dpanagop@biophysics.gr

channels on the cell membranes according to the Ion Forced-Vibration mechanism that we have previously proposed. This biophysical mechanism seems to be realistic, since it is able to explain all of the reported biological effects associated with electromagnetic fields (EMFs) exposure, including the so-called "windows" of increased bioactivity reported since many years but remained unexplained so far, and recorded also in our recent experiments in regards to GSM radiation exposure. The chapter discusses also, an important dosimetry issue, regarding the use of Specific Absorption Rate (SAR), a quantity introduced to describe temperature increases within biological tissue (thermal effects), while the vast majority of the recorded biological effects are non-thermal. Finally the chapter attempts to propose some basic precautions and a different way of mobile telephony base station antennas network design, in order to minimize the exposure of human population and reduce significantly the current exposure limits in order to account for the reported non thermal biological effects.

Keywords: mobile telephony radiation, GSM, RF, ELF, electromagnetic fields, non-ionizing electromagnetic radiation, biological effects, health effects, Drosophila, reproductive capacity, DNA damage, cell death, intensity windows, SAR.

INTRODUCTION

MobileTelephony Radiation such as GSM and 3G (3rd generation) (Curwen and Whalley 2008) is probably the main source of public microwave exposure in our time. Billions of people globally are self-exposed daily by their own mobile phones, while at the same time they are also exposed by base station antennas which are installed within residential and working areas. While exposure from mobile phones is voluntary for every user for as long daily periods as each one decides, exposure from base station antennas -although weaker- is involuntary and constant for up to 24 h a day.

A large number of biological, clinical and epidemiological studies regarding the possible health and environmental implications of microwave exposure is already published, (for a review see Panagopoulos and Margaritis 2008; 2009; 2010a). While many of these studies do not report any effect, many others are indicating serious biological, clinical and health effects such as DNA damage, cell death, reproductive decreases, sleep disturbances, electroencephalogram (EEG) alterations, and cancer induction.

Some of the studies report DNA damage or cell death or oxidative stress induction on reproductive insect and mammalian (including human) cells (Panagopoulos et al. 2007a; 2010; De Iuliis et al. 2009; Agarwal et al. 2009; Mailankot et al. 2009; Yan et al. 2007). The findings of these studies seem to explain the results of other studies that simply report insect, bird, and mammalian (including human) infertility (Panagopoulos et al. 2004; 2007b; Gul et al. 2009; Agarwal et al. 2008; Batellier et al. 2008; Wdowiak et al. 2007; Magras ans Xenos 1997). Other recent reports regarding reduction of insect (especially bees) and bird populations during the last years (Stindl and Stindl 2010; Bacandritsos et al. 2010; van Engelsdorp et al. 2008; Everaert and Bauwens 2007; Balmori 2005), also seem to correlate with the above mentioned studies since their findings may be explained by cell death induction on reproductive cells. Other studies report DNA damage or oxidative stress induction or increase in cellular damage features in somatic mammalian and insect cells after *in vitro* or *in vivo* exposure to microwaves (Guler et al. 2010; Tomruk et al. 2010; Franzellitti et al. 2010; Luukkonen et al. 2009; Yao et al. 2008; Yadav and Sharma 2008; Sokolovic et al. 2008; Lee et al. 2008; Lixia et al. 2006; Zhang et al. 2006; Nikolova et al. 2005; Belyaev et al. 2005; Diem et al. 2005). At the same time, some other studies report brain tumour induction in humans, (Hardell et al. 2009; 2007; Khurana et al. 2009; Johansson 2009), or symptoms of unwellness among people residing around base station antennas (Hutter et al. 2006; Salama et al. 2004; Navarro et al. 2003).

In spite of many other studies that report no effects (see Panagopoulos and Margaritis 2008; 2009; 2010a), the similarity of the above findings between them and their rapidly increasing number during the last years is of great importance. All the above-mentioned recent studies from different research groups and on different biological models exhibit mutually supportive results and this makes unlikely the possibility that these results could be either wrong or due to randomness. While recent experimental findings tend to show a distinct similarity between them, the need for a biophysical and biochemical explanation on the basis of a realistic mechanism of action of EMFs at the cellular level, becomes more and more demanding.

While there is still no widely accepted biophysical or biochemical mechanism to explain the above findings at cellular level, many recent findings tend to support the possibility that oxidative stress and free radical action may be responsible for the recorded genotoxic effects of EMFs which may lead to health implications and cancer induction. It is possible that free radical action and/or irregular release of hydrolytic enzymes like DNases, induced by EMFs exposure, may constitute the biochemical action leading to DNA damage. This biochemical action may be initiated by alterations in intracellular ionic concentrations after irregular gating of electro-sensitive channels on cell membranes by external EMFs. Such irregular gating of ionic channels may represent the more fundamental biophysical mechanism to initiate the biochemical one, as previously supported by us (Panagopoulos et al. 2000; 2002).

EFFECTS OF MOBILE TELEPHONY RADIATION ON A MODEL BIOLOGICAL SYSTEM

We shall describe here the effects of the two systems of Mobile Telephony radiation used in Europe, GSM 900 MHz and GSM 1800 MHz (also called DCS –Digital Cellular System), on a model biological system, the reproductive capacity of the insect *Drosophila melanogaster*.

The reproductive capacity of animals depends on their ability to successfully complete subtle biological functions, as is gametogenesis (oogenesis, spermatogenesis), fertilization, and embryogenesis, in spite of any disturbing exogenous (or endogenous) factors. In the experiments that will be presented here the exogenous disturbing factor is the Radio-Frequency (RF)/microwave radiation-fields used in modern mobile telecommunications.

Gonad development (oogenesis, spermatogenesis) in all animals is a biological process, much more sensitive to environmental stress than other developmental - biological processes that take place at later stages of animal development. This is shown in relation to ionizing radiation as stress factor, it is in agreement with the empirical law of Bergonie-Tribondau (Coggle 1983; Hall and Giaccia 2006) and it is verified also in relation to non-ionizing radiation by several recent experimental results, including the following ones.

The reproductive capacity of *Drosophila melanogaster* (especially oogenesis and spermatogenesis) is a model biological system, very well studied with a very good timing of its developmental processes under certain laboratory conditions (King 1970; Panagopoulos et al. 2004; Horne-Badovinac and Bilder 2005).

Following a well tested protocol of ours, the reproductive capacity is defined by the number of F_1 (first filial generation) pupae, which under the conditions of our experiments corresponds to the number of laid eggs (oviposition), since there is no statistically significant mortality of fertilized eggs, larvae or pupae derived from newly eclosed adult flies during the first days of their maximum oviposition (Panagopoulos et al. 2004).

Basic Experimental Procedure

All sets of experiments were performed with the use of commercially available cellular mobile phones as exposure devices.

The exposures were performed with the mobile phone antenna outside of the glass vials containing the flies, in contact with, or at certain distances from the glass walls. The daily exposure duration was a few minutes (depending on the kind of experiments – see below), in one dose. The exposures always started on the first day (day of eclosion) of each experiment, and lasted for a total of five or six days.

The temperature during the exposures was monitored within the vials by a mercury thermometer with an accuracy of $0.05°C$ (Panagopoulos et al. 2004).

In each experiment, we collected newly emerged adult flies from the stock; we anesthetized them very lightly and separated males from females. We put the collected flies in groups of ten males and ten females in standard laboratory 50-ml cylindrical glass vials (tubes), with 2.5cm diameter and 10cm height, with standard food, which formed a smooth plane surface 1cm thick at the bottom of the vials. The glass vials were closed with cotton plugs.

In each group we kept the ten males and the ten females for the first 48h of the experiment in separate glass vials. Keeping males separately from females for the first 48h of the experiment ensures that the flies are in complete sexual maturity and ready for immediate mating and laying of fertilized eggs, (Panagopoulos et al. 2004).

After the first 48h of each experiment, the males and females of each group were put together (ten pairs) in another glass tube with fresh food, allowed to mate and lay eggs for 72h. During these three days, the daily egg production of *Drosophila* is at its maximum.

After five days from the beginning of each experiment the flies were removed from the glass vials and the vials were maintained in the culture room for six additional days, without any further exposure to the radiation. The removed maternal flies depending on each separate experimental series, could be collected and their ovaries were dissected and treated for different biochemical assays (see below).

After the last six days, most F_1 embryos (deriving from the laid eggs) are in the stage of pupation, where they can be clearly seen with bare eyes and easily counted on the walls of the glass tubes.

We have previously shown that this number of F_1 pupae is a representative estimate of the insect's reproductive capacity (Panagopoulos et al. 2004).

Exposures and measurements of mobile phone emissions were performed at the same place within the lab, where the mobile phone had full reception of the GSM signals.

1. COMPARISON OF BIOACTIVITY BETWEEN NON-MODULATED (DTX) AND MODULATED (TALK SIGNAL) GSM RADIATION

In the first series of experiments, (parts 1A and 1B) we separated the insects into two groups: a) the Exposed group (E) and b) the Sham Exposed (Control) group (SE). Each one of the two groups consisted of ten female and ten male, newly emerged adult flies. The sham exposed groups had identical treatment as the exposed ones, except that the mobile phone during the "exposures" was turned off.

The total duration of exposure was 6min per day in one dose and we started the exposures on the first day of each experiment (day of eclosion). The exposures took place for a total of 5 days.

In the first part of these experiments (1A) the insects were exposed to Non-Modulated GSM 900 MHz radiation (TDX -discontinuous transmission mode-signal) while in the second part (1B) they were exposed to Modulated GSM 900 MHz radiation (or "GSM talk signal"). In both cases, the exposures were performed with the antenna of the mobile phone in contact with the walls of the glass vials containing the insects.

The difference between the modulated and the corresponding non-modulated GSM radiation is that, the intensity of the modulated radiation is about ten times higher than the intensity of the corresponding non-modulated from the same handset (mobile phone) and additionally, the modulated radiation embodies more and larger variations in its intensity within the same time interval, than the corresponding non-modulated (Panagopoulos et al. 2004; Panagopoulos and Margaritis 2008).

The mean power density for 6 min of Modulated emission, with the antenna of the mobile phone outside of the glass vial in contact with the glass wall and parallel to the vial's axis, was 0.436±0.060 mW/cm^2 and the corresponding mean value for Non-Modulated (NM) emission, 0.041±0.006 mW/cm^2. The measured Extremely Low Frequency (ELF) mean values of electric field intensity of the GSM signals excluding the ambient fields of 50Hz, were 6.05±1.62 V/m for the Modulated signal, and 3.18±1.10 V/m for the Non-Modulated signal. These values are averages from eight separate measurements of each kind ± Standard Deviation (SD).

1A. Experiments with Non-Modulated GSM 900 MHz radiation ("non-speaking" emission or DTX-signal), showed that this radiation decreases insect reproduction by an average of 18.24 %, after 6 min daily exposure for 5 consecutive days (Table 1).

The exposure conditions in these experiments simulate the potential biological impact on a mobile phone user who listens through the mobile phone during a conversation, with the handset close to his/her head.

The average mean numbers of F_1 pupae from 4 identical separate experiments (corresponding to the number of laid eggs) per maternal fly in the groups E(NM) exposed to Non-Modulated (NM) GSM radiation-field, and in the corresponding sham exposed (control) groups SE(NM) during the first three days of the insect's maximum oviposition, are shown in the first two rows of Table 1.

Statistical analysis, (single factor ANOVA test) showed that the probability that differences between the groups exposed to non-modulated GSM radiation and the sham exposed groups, owing to random variations, is $P < 5 \times 10^{-4}$, meaning that, the decrease in the reproductive capacity is actually due to the effect of the GSM signal. [A detailed description of these experiments can be found in Panagopoulos et al. 2004].

1B. Experiments with Modulated GSM 900 MHz radiation ("speaking" emission or "GSM Talk signal") exposure, showed that this radiation decreases insect reproduction by an average of 53.01 %, after 6 min daily exposure for 5 consecutive days (Table 1).

The experimenter spoke close to the mobile phone's mic during the exposures. The exposure conditions in this case simulate the potential biological impact when a user speaks on the mobile phone during a conversation, with the handset close to his/her head.

The last two rows of Table 1 show the average mean number of F_1 pupae from 4 identical separate experiments (corresponding to the number of laid eggs) per maternal fly in the groups E(M), exposed to "Modulated" (M) GSM radiation- field and in the corresponding sham exposed groups, SE(M), during the first three days of the insect's maximum oviposition.

Table 1. Effect of Non-Modulated (DTX) and Modulated (Talk mode) GSM Radiation on the Reproductive Capacity of Drosophila melanogaster

Type of GSM 900 MHz Radiation	Groups	Average Mean Number of F_1 Pupae per Maternal Fly in four separate experiments ± SD	Deviation from corresponding Control	Probability that Differences between Exposed and corresponding Sham-Exposed Groups are due to random variations
NM or DTX-signal	E(NM)	9.97 ± 0.31	-18.24%	$P < 5 \times 10^{-4}$
	SE(NM)	12.2 ± 0.57		
M or Talk-signal	E(M)	5.85 ± 0.67	-53.01%	$P < 10^{-5}$
	SE(M)	12.45 ± 0.6		

The statistical analysis shows that the probability that mean oviposition differs between the groups exposed to modulated GSM radiation and the sham-exposed groups, owing to random variations, is very small, $P < 10^{-5}$. Thus the recorded effect is actually due to the GSM signal.

Although the intensity of the modulated signal is about ten times higher than the corresponding intensity of the non-modulated signal, the reproductive capacity was decreased by 53.01 % by the modulated emission, and 18.24 % by the non-modulated one. Thus the effect seems to be strongly, but non-linearly, dependent on the radiation intensity.

The results from the first set of experiments (parts 1A and 1B) are graphically represented, in Figure 1.

Temperature increases were not detected within the vials during the 6 min exposures with either DTX or "Talk signal". Therefore the described effects are considered as non-thermal.

Bioactivity of Non-Modulated and Modulated GSM Radiation

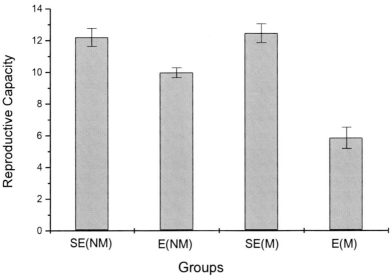

Figure 1. Reproductive Capacity (average mean number of F1 pupae per maternal insect) ± SD of the insect groups exposed to non-modulated and modulated GSM 900 MHz radiation [E(NM), E(M)] and the corresponding sham-exposed, [SE(NM), SE(M)], groups.

2. EFFECT OF GSM RADIATION ON MALES AND FEMALES

In this set of experiments, the effect of GSM 900 MHz field on the reproductive capacity of each sex separately was investigated. The mobile phone was operating in speaking mode during the 6 min exposures, and the insects were separated into four groups (each one consisting again of 10 male and 10 female insects): In the first group (E1) both male and female insects were exposed. In the second group (E2) only the females were exposed. In the third group (E3) only the males were exposed and, the fourth group (SE) was sham-exposed (control). Therefore in this set of experiments, the 6-min daily exposures took place only during the first two days of each experiment while the males and females of each group were separated, and the total number of exposures in each experiment was 2 instead of 5. The exposures were again performed with the antenna of the mobile phone in contact with the glass vials containing the insects.

The average mean number of F_1 pupae per maternal fly from 4 identical separate experiments, of all the groups, are given in Table 2 and represented graphically in Figure 2.

The statistical analysis (single factor ANOVA test) shows that the probability that the mean number of F_1 pupae differs between the four groups because of random variations is, $P < 10^{-7}$.

These results show that the GSM radiation-field decreases the reproductive capacity of both female and male insects. The reason why female insects (E2) appear to be more affected than males (E3), is probably that, by the time we started the exposures, spermatogenesis was already almost completed in male flies, while oogenesis had just started (King 1970;

Panagopoulos et al. 2004). Therefore it should be expected that the GSM exposure would affect oogenesis more than spermatogenesis and the decrease in reproductive capacity would be more evident in the female than in the male insects.

Table 2. Effect of "Modulated" GSM Radiation on the Reproductive Capacity of each Sex

Groups	Average Mean Number of F_1 Pupae per Maternal Fly ± SD	Deviation from Control
E1	7.7 ± 0.66	-42.32%
E2	8.85 ± 0.73	-33.71%
E3	11.75 ± 0.54	-11.985%
SE (Control)	13.35 ± 0.39	

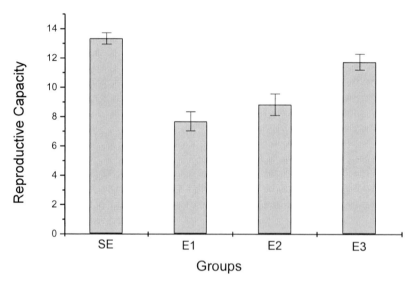

Figure 2. Effect of Modulated GSM radiation on the reproductive capacity of each sex of Drosophila melanogaster. Average mean number of F1 pupae per maternal insect ± SD. SE: sham exposed groups, E1: groups in which both sexes were exposed, E2: groups in which only the females were exposed, E3: groups in which only the males were exposed.

3. COMPARISON OF BIOACTIVITY BETWEEN GSM 900 AND GSM 1800

GSM 900 MHz antennas of both handsets and base stations operate at double the power output than the corresponding GSM 1800 MHz antennas. [As mentioned before, GSM 1800 MHz radiation is also referred to as DCS]. Additionally, the two systems use different carrier frequencies (900 or 1800 MHz respectively). Therefore, a comparison of the biological activity between the two European systems of Mobile Telephony radiation is of great importance. [GSM 1900 MHz system operating in the USA, is similar to GSM 1800 MHz, except for the 100 MHz difference of their carrier frequencies].

In this and the next series of experiments, we used a dual band cellular mobile phone that could be connected to either GSM 900 or 1800 networks, simply by changing SIM ("Subscriber Identity Module") cards on the same handset. The highest Specific Absorption Rate (SAR), given by the manufacturer for human head, was 0.89 W/Kg. The exposure procedure was the same. The handset was fully charged before each set of exposures. The experimenter spoke on the mobile phone's microphone during the exposures. The GSM 900 and 1800 fields were thus "modulated" by the human voice, ("speaking emissions" or "GSM talk signals").

The exposures and the measurements of the mobile phone emissions were always performed at the same place within the lab, where the mobile phone had full reception of both GSM 900 and 1800 signals.

The measured mean power densities in contact with the mobile phone antenna for six min of modulated emission, were 0.407 ± 0.061 mW/cm^2 for GSM 900 MHz and 0.283 ± 0.043 mW/cm^2 for GSM 1800 MHz. As expected, GSM 900 MHz intensity at the same distance from the antenna and with the same handset was higher than the corresponding 1800 MHz. For a better comparison between the two systems of radiation we measured the GSM power density at different distances from the antenna and found that at 1cm distance, the GSM 900 MHz intensity was 0.286 ± 0.050 mW/cm^2, almost equal to GSM 1800 MHz at zero distance. Measured electric and magnetic field intensities in the ELF range for the modulated field, excluding the ambient electric and magnetic fields of 50Hz, were 22.3 ± 2.2 V/m electric field intensity and 0.50 ± 0.08 mG magnetic field intensity for GSM 900 at zero distance, 13.9 ± 1.6 V/m, 0.40 ± 0.07 mG correspondingly for GSM 900 at 1 cm distance and 14.2 ± 1.7 V/m, 0.38 ± 0.07 mG correspondingly for GSM 1800 at zero distance. All these values are averaged over ten separate measurements of each kind ± standard deviation (SD).

Each type of radiation gives a unique frequency spectrum. While GSM 900 MHz gives a single peak around 900 MHz, GSM 1800 MHz gives a main peak around 1800 MHz and a smaller one around 900 MHz, (Panagopoulos et al. 2007b).

In this set of experiments we separated the insects into four groups: a) the group exposed to GSM 900 MHz field with the mobile phone antenna in contact with the glass vial containing the flies (named "900"), b) the group exposed to GSM 900 MHz field with the antenna of the mobile phone at 1cm distance from the vial (named "900A"), c) the group exposed to GSM 1800MHz field with the mobile phone antenna in contact with the glass vial (named "1800"), and d) the sham-exposed (control) group (named "SE"). The comparison between the first and the third groups represents comparison of potential biological impact between GSM 900 and GSM 1800 users under the actual exposure conditions. Comparison between the first and the second groups represents comparison of bioactivity between signals of different intensity but of the same carrier frequency, and finally, comparison between the second and the third groups represents comparison of bioactivity between the RF carrier frequencies of the two systems under equal radiation intensities. Therefore the second group (900A) was introduced for better comparison of the effects between the two types of radiation.

The average mean numbers of F_1 pupae from ten replicate experiments for the different groups, are given in Table 3 and represented graphically, in Figure 3.

The results from this set of experiments show that the reproductive capacity in all the exposed groups is significantly decreased compared to the sham exposed groups. The average decrease in ten replicate experiments was found to be maximum in the 900 groups, (48.25% compared to SE) and smaller in the 900A and the 1800 groups (32.75% and 31.08% respectively), (Table 3). Although the decrease was even smaller in the 1800 than in 900A groups, differences between the 900A and 1800 groups were found to be within the standard deviation, (Table 3, Figure 3).

Table 3. Effect of Modulated GSM 900 and GSM 1800 fields on the Reproductive Capacity of Drosophila melanogaster

Groups	Average Mean Number of F_1 Pupae per Maternal Fly in ten replicate experiments ± SD	Deviation from Control
900	6.51 ± 0.67	-48.25%
900A	8.46 ± 0.55	-32.75%
1800	8.67 ± 0.65	-31.08%
SE (Control)	12.58 ± 0.95	

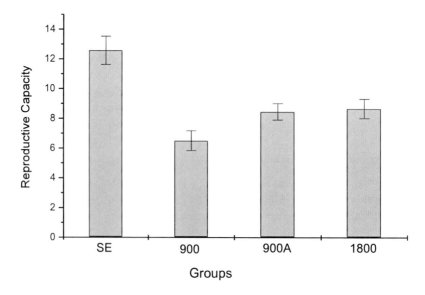

Figure 3. Reproductive Capacity (average mean number of F_1 pupae per maternal insect) ± SD of insect groups exposed to GSM 900 and GSM 1800 radiations (900, 900A, 1800) and sham-exposed (SE) groups.

The statistical analysis showed that the probability that the reproductive capacity differs between the four groups, owing to random variations, was negligible, $P < 10^{-18}$.

Temperature increases were again not detected within the glass vials during the exposures.

The differences in the reproductive capacity between the groups were larger between 900 and 900A (owing to intensity differences between the two types of radiation) and much smaller between 900A and 1800, (owing to frequency difference of the carrier signal between GSM 900 and 1800), (Table 3).

This set of experiments showed that there is a difference in the bioactivity between GSM 900 MHz and GSM 1800 MHz. The GSM 900 signal is more bioactive than the corresponding GSM 1800 signal under equal other conditions and the difference is mostly due to the higher intensity of GSM 900 under the same exposure conditions, (differences between groups 900 and 900A) and less due to the different RF carrier frequencies, (differences between 900A and 1800 groups).

Intensity differences between the two types of cellular mobile telephony radiation depend also on the ability of communication between the antennas of the mobile phone and the corresponding base station. Even if GSM 900 usually has a higher intensity than GSM 1800, this situation can be reversed in certain places where GSM 900 has a better signal reception between its antennas than GSM 1800. This is because when GSM antennas of both systems cannot easily communicate between base station and mobile phone, they emit stronger signals in order to achieve communication. Our results count for equal signal reception conditions between the two types of radiation.

A detailed description of these experiments can be found in Panagopoulos et al. 2007b.

4. GSM BIOACTIVITY ACCORDING TO ITS INTENSITY (OR ACCORDING TO THE DISTANCE FROM THE ANTENNA). THE REVELATION OF A "WINDOW" OF INCREASED BIOACTIVITY

Until the recent publication of this set of experiments (Panagopoulos et al. 2010; Panagopoulos and Margaritis 2008; 2009), no other experiments were reported regarding the effects at different distances from mobile phone antennas corresponding to different intensities of the emitted radiation, neither experiments regarding the effects of mobile telephony base station antennas, except of statistical observations which had reported a reduction of bird populations around base station antennas (Everaert and Bauwens 2007; Balmori 2005).

Radiation from base station antennas is almost identical to that from mobile phones of the same system (GSM 900 or 1800), except that it is about 100 times more powerful, and uses a little higher carrier frequency. GSM 900 mobile phones emit between 890 MHz and 915 MHz (uplink operation) while base stations emit between 935 MHz and 960 MHz (downlink operation). The corresponding GSM 1800 spectrums are 1710-1785 MHz (uplink operation) and 1805-1880 MHz (downlink operation). Another difference is that, although the time-averaged emitted power is significantly higher in base station antennas than in the mobile phones, the ratio of pulse peak power versus time-averaged power is higher in the mobile phones (Hillebrand 2002; Clark 2001; Hyland 2000; Hamnerius and Uddmar 2000; Tisal 1998; Panagopoulos and Margaritis 2008). Still, the two kinds of radiation are very similar and effects produced by mobile phones at certain distances, can be extrapolated to represent effects from base station antennas, of the same type of radiation, at about 100 times longer distances. Thus, distances from mobile phone antennas can be corresponded to about 100 times longer distances from base station antennas of the same type of radiation. For example, when our distance from a mobile phone during connection is 2 m, (e.g. someone talking on the mobile phone at 2 m distance from us), then we are exposed almost equally as by a corresponding base station antenna at 200 m distance. Correspondingly, if a mobile telephony base station antenna is installed at 200 m from our residential place, this is almost the same as when we are exposed by a mobile phone operating in talk mode 24 h a day at 2 m distance from us.

The difficulty in performing experiments with base station mobile telephony antennas is due to the fact of uncontrolled conditions in the open air that do not allow the use of sham-exposed animals, (exposed to identical other conditions like temperature, humidity, light etc). In other words, there is no way to have a sham-exposed group of experimental animals under identical environmental conditions as the exposed ones, but without being exposed to the radiation at the same time. The only way to simulate the reality of the exposure by a base station antenna was to expose the animals at different distances from a mobile phone within the lab.

In order to study the bioactivity of mobile telephony signals at different intensities-distances from the antenna of a mobile phone handset, resembling effects of base station signals within residential areas, we used the same biological index, the reproductive capacity of the insect *Drosophila melanogaster*, defined by the number of F_1 pupae derived during the three days of the insect's maximum oviposition.

In each experiment of this set, we separated the collected insects into thirteen groups: The first group (named "0") was exposed to GSM 900 or 1800 field with the mobile phone antenna in contact with the glass vial containing the flies. The second (named "1"), was exposed to GSM 900 or 1800 field, at 1cm distance from the mobile phone antenna. The third group (named "10") was exposed to GSM 900 or 1800 field at 10 cm distance from the mobile phone antenna. The fourth group (named "20") was exposed to GSM 900 or 1800 field at 20 cm distance from the mobile phone antenna, etc, the twelfth group (named "100") was exposed to GSM 900 or 1800 field at 100 cm distance from the mobile phone antenna. Finally, the thirteenth group (named "SE") was the sham exposed. Each group consisted of ten male and ten female insects as always.

Radiation and field measurements in contact and at different distances from the mobile phone antenna, for six min of modulated emission, for GSM 900 MHz and 1800 MHz in the RF and ELF ranges excluding the background electric and magnetic fields of 50 Hz, are given in Table 4. All values are averaged over ten separate measurements of each kind ± standard deviation (SD). The measurements reveal that although ELF electric and magnetic fields, associated with the GSM signals, fall within the background of the 50 Hz electric/magnetic power transmission fields for distances longer than 50 cm from both GSM 900 and GSM 1800 mobile phone antennas and for this cannot be detected, the RF components of the signals are still evident for distances up to 100 cm (Table 4).

It is important to clarify that, the fact that the ELF components of the GSM signals fall within the background levels, does not mean that they do not exist. On the contrary, they may still be bioactive even though they cannot be easily detected.

Table 4. GSM 900 and 1800 Radiation and Field Intensities ± SD, in the Microwave and ELF regions, for different Distances from a mobile phone Antenna (1)

Distance from mobile phone Antenna (cm)	GSM 900 Radiation Intensity at 900 MHz, (mW/cm²)	GSM 900 Electric Field Intensity at 217 Hz, (V/m)	GSM 900 Magnetic Field Intensity at 217 Hz, (mG)	GSM 1800 Radiation Intensity at 1800 MHz, (mW/cm²)	GSM 1800 Electric Field Intensity at 217 Hz, (V/m)	GSM 1800 Magnetic Field Intensity at 217 Hz, (mG)
0	0.378 ±0.059	19 ±2.5	0.9 ±0.15	0.252 ±0.050	13 ±2.1	0.6 ±0.08
1	0.262 ±0.046	12 ±1.7	0.7 ±0.13	0.065 ±0.015	6 ±0.8	0.4 ±0.07
10	0.062 ±0.020	7 ±0.8	0.3 ±0.05	0.029 ±0.005	2.7 ±0.5	0.2 ±0.05
20	0.032 ±0.008	2.8±0.4	0.2 ±0.04	0.011 ±0.003	0.6 ±0.12	0.1±0.02
30	0.010 ±0.002	0.7 ±0.09	0.1 ±0.02	0.007 ±0.001	0.3 ±0.06	0.06 ±0.01
40	0.006 ±0.001	0.2 ±0.03	0.05 ±0.01	0.004 ±0.0007	0.1 ±0.04	-
50	0.004 ±0.0006	0.1 ±0.02	-	0.002 ±0.0003	-	-
60	0.002 ±0.0003	-	-	0.0016 ±0.0002	-	-
70	0.0017 ±0.0002	-	-	0.0013 ±0.0002	-	-
80	0.0012 ±0.0002	-	-	0.0011 ±0.0002	-	-
90	0.0010 ±0.0001	-	-	0.0005 ±0.0001	-	-
100	0.0004 ±0.0001	-	-	0.0002 ±0.0001	-	-

(1) For distances longer than 30-50 cm from the mobile phone antenna, the ELF electric and magnetic field components of both GSM 900 and 1800 radiations, fall within the background of the stray 50 Hz fields within the lab.

In each experiment all the 12 exposed groups were simultaneously exposed during the 6 min exposure sessions. After each exposure, the corresponding sham-exposure took place. The SE group was "exposed" for 6 min at zero distance from the mobile phone antenna, following exactly the same methodology (the experimenter spoke on the mobile phone, same voice, reading the same text) but the mobile phone was turned off. It was already verified by preliminary experiments, that SE groups at all the 12 different locations of exposure, did not differ significantly between them in their reproductive capacity.

The average mean values of reproductive capacity (mean number of F_1 pupae per maternal insect) from eight separate identical experiments with GSM 900 and GSM 1800 exposures are listed in Table 5 and represented graphically in Figures 4 and 5.

The data show that GSM 900 mobile telephony radiation decreases reproductive capacity at distances from 0 cm up to 90 cm from the mobile phone antenna, (corresponding intensities ranging from 378 µW/cm² down to 1 µW/cm² – Tables 4, 5). Table 5 and Fig 4 show that the effect is at a maximum at 0 cm and at 30 cm from the antenna, (corresponding to radiation intensities of 378 µW/cm² and 10 µW/cm² respectively) with overall maximum at 30 cm. For distances longer than 30 cm from the mobile phone antenna, the effect decreases rapidly and becomes very small for distances longer than 50 cm, but it is still evident for distances up to 90 cm (intensities down to 1 µW/cm²).

Table 5. Effect of GSM 900 and 1800 radiation-fields on Reproductive Capacity at different Distances from the Antenna

Groups - Distance from mobile phone antenna, (cm)	Average Mean Number of F_1 Pupae per Maternal Fly± SD, for GSM 900 MHz	Deviation from Sham Exposed Group	Average Mean Number of F_1 Pupae per Maternal Fly± SD, for GSM 1800 MHz	Deviation from Sham Exposed Group
0	7.46 ± 0.73	-46.14 %	9.10 ± 0.69	-35.09 %
1	9.35 ± 0.62	-32.49 %	11.35 ± 0.63	-19.04 %
10	11.28 ± 0.81	-18.56 %	11.93 ± 0.72	-14.91 %
20	11.55 ± 0.79	-16.61 %	8.33 ± 0.7	-40.58 %
30	7.38 ± 0.65	-46.71 %	12.77 ± 0.82	-8.92 %
40	12.81 ± 0.97	-7.51 %	13.52 ± 0.86	-3.57 %
50	13.49 ± 0.82	-2.60 %	13.72 ± 0.75	-2.14 %
60	13.62 ± 0.83	-1.66 %	13.81 ± 0.92	-1.50 %
70	13.72 ± 0.92	-0.94 %	13.79 ± 0.90	-1.64 %
80	13.68 ± 0.80	-1.23 %	13.85 ± 0.81	-1.21 %
90	13.75± 0.95	-0.72 %	14.03 ± 1.02	+0.07 %
100	14.01 ± 1.01	+1.16 %	14.05 ± 0.99	+0.21 %
SE	13.85 ± 0.91		14.02 ± 0.98	

The data also show that GSM 1800 mobile telephony radiation decreases reproductive capacity at distances from 0 cm up to 80 cm from the mobile phone antenna, (corresponding intensities ranging from 252 µW/cm² down to 1.1 µW/cm² – Tables 4, 5). Table 5 and Fig. 5 show that the effect is maximum at 0 cm and at 20 cm from the antenna, (corresponding to radiation intensities of 252 µW/cm² and 11 µW/cm² respectively) with overall maximum at

20 cm. For distances longer than 20 cm from the mobile phone antenna, the effect decreases rapidly and becomes very small for distances longer than 40 cm, but it is still evident for distances up to 80 cm (intensities down to 1.1 µW/cm^2).

Thus, the effect of mobile telephony radiation on reproductive capacity is at a maximum at zero distance (intensities higher than 250 µW/cm^2) and then becomes maximum at a distance of 30 cm or 20 cm from the mobile phone antenna for GSM 900 or 1800 MHz radiation respectively. These distances of 30 cm and 20 cm respectively, correspond to the same RF intensity of about 10 µW/cm^2 and also to the same ELF electric field intensity of about 0.6-0.7 V/m (Table 4).

Bioactivity of GSM 900 according to Distance from the Antenna

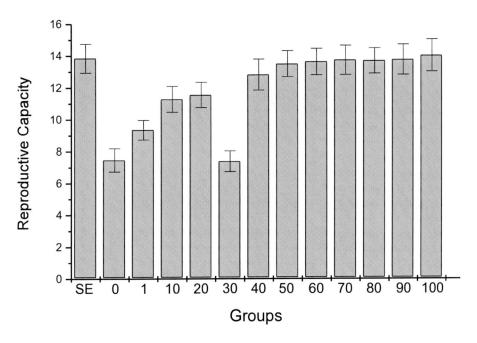

Figure 4. Reproductive Capacity (average mean number of F1 pupae per maternal insect) ± SD in relation to the Distance from a GSM 900 MHz mobile phone antenna. The decrease in reproductive capacity is maximum at zero distance and at 30 cm distance from the antenna ("window" of increased bioactivity), corresponding to RF intensities 378 µW/cm2 and 10 µW/cm2, (Tables 4, 5).

Again, there were no temperature increases within the vials during the exposures at all the different distances from the mobile phone handset.

The effect diminishes considerably for distances longer than 50 cm from the mobile phone antenna and disappears for distances longer than 80-90 cm, corresponding to radiation intensities smaller than 1 µW/cm^2. For distances longer than 50 cm where the ELF components fall within the background, the decrease in reproductive capacity is within the standard deviation. This might suggest that the ELF components of digital mobile telephony signals, play a key role in their bio-activity, alone or in conjunction with the RF carrier signal.

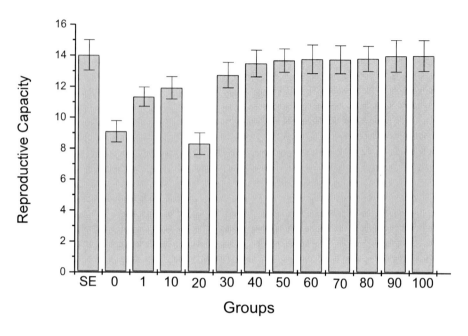

Figure 5. Reproductive Capacity (average mean number of F1 pupae per maternal insect) ± SD in relation to the Distance from a GSM 1800 MHz mobile phone antenna. The decrease in reproductive capacity is maximum at zero distance and at 20 cm distance from the antenna ("window" of increased bioactivity), corresponding to RF intensities 252 µW/cm2 and 11 µW/cm2, (Tables 4, 5).

The results on reproductive capacity were analyzed statistically by single factor Analysis of Variance. In addition, linear (Pearson's) and non-parametric (Kendall's) correlation analysis were performed between reproductive capacity and radiation/field intensities in order to get an estimation of which parameter (RF radiation or ELF fields) might be more responsible for the effects (Weiss 1995; Maber 1999).

Single factor Analysis of Variance test, showed that the probability that the reproductive capacity differs between all groups, owing to random variations, is negligible both for GSM 900 and 1800 exposures, $P < 10^{-27}$ in both cases. The results of (Pearson's) linear correlation analysis show a slightly stronger linear relationship between reproductive capacity and ELF electric field intensity, (linear correlation coefficient, $r \cong -0.72$, $P<0.01$ for GSM 900 and $r \cong -0.65$, $P<0.03$ for GSM 1800), than between reproductive capacity and RF radiation intensity ($r \cong -0.70$, $P<0.02$ and $r \cong -0.63$, $P<0.03$ respectively), both for GSM 900 and 1800 exposures. Since our results show that the dependence of reproductive capacity on RF and ELF intensities is non-linear, (Fig. 4, 5), we applied also Kendall's non-parametric correlation analysis for a better estimation of the non-linear correlation between the variables. This correlation analysis in contrast to the previous one, showed a slightly stronger relationship between reproductive capacity and RF radiation intensity (correlation coefficient, $r \cong -0.85$, $P<0.001$ for GSM 900 and $r \cong -0.88$, $P<0.001$ for GSM 1800), than between reproductive capacity and ELF electric field intensity $r \cong -0.79$, $P = 0.001$ and $r \cong -0.78$, $P = 0.001$ respectively), both for GSM 900 and 1800 exposures. We note that the P-values (the probabilities that the corresponding *r*-values are due to random variation in the data points) in the case of Kendall's non-parametric correlation are smaller than the corresponding ones in Pearson's linear correlation, suggesting that non-parametric correlation analysis is perhaps more appropriate in the case of our (non-linear) results. The correlation analysis between reproductive capacity and distance from the antenna, gave the same values as between reproductive capacity and RF intensity and the correlation between reproductive capacity and ELF magnetic field was found to be even weaker than between reproductive capacity and ELF electric field.

It is interesting that the decrease in the reproductive capacity was found to be maximum not only within the near field of the mobile phone antenna (0-5.2 cm from the antenna for GSM 900 and 0-2.6 cm for GSM 1800) (Panagopoulos and Margaritis 2010b), where the intensity of the radiation is maximum, but also within the far field, at 20-30 cm distance from the mobile phone antenna, where the intensity is significantly decreased.

Thus, we have discovered the existence of increased bioactivity "windows" for both GSM 900 and 1800 radiations. These "bioactivity windows" appear at distances 20 cm and 30 cm from the GSM 1800 and 900 mobile phone antennas respectively, where the radiation intensity is in both cases close to 10 $\mu W/cm^2$ and the ELF electric field intensity $0.6 - 0.7$ V/m. At these distances, the bio-effect becomes even more intense than at zero distance from a mobile phone antenna where the RF intensity is higher than 250 $\mu W/cm^2$, and the ELF electric filed intensity higher than 13 V/m (Table 4).

The distance of 20-30 cm from the mobile phone antenna, at which the windows of increased bioactivity appear, corresponds to a distance of about 20-30 meters from a base station antenna. Since mobile telephony base station antennas are most usually located within residential areas, at distances 20-30 m from such antennas there are often houses and work places where people are exposed for up to 24 hours per day. Therefore the existence of these "windows" may pose an increased danger for people who reside or work at such distances from mobile telephony base station antennas. Our present findings show that mobile telephony radiation can be very bioactive at intensity levels encountered at residential and working areas around base station antennas.

From these results, it became evident that another series of experiments was necessary, aiming to reveal the nature of these bioactivity "windows", (i.e. whether they depend on the intensity of the radiation/fields, or on any other parameter like for example the wavelength of the radiation which happens to be close to the distance where the "window" appears – 17 cm approximately for 1800 MHz and 33 cm approximately for 900 MHz), (Panagopoulos and Margaritis 2010b).

We do not know which constituent of the real mobile telephony signal, (i.e. the RF carrier, the ELF pulse repetition frequencies, or the combination of both), is more responsible for the bioactivity of the signal or for the existence of the "windows" found in our experiments. Real mobile telephony signals are always RF carrier signals pulsed at ELF in order to be able to transmit information. Furthermore, real mobile telephony signals are never constant in intensity or frequency. Therefore, experiments with idealized continuous signals corresponding to the RF carrier alone or to the ELF constituents alone, do not represent real conditions.

The fact that for distances longer than 50 cm where the ELF components fall within the background, the bioactivity of the radiation, although still evident, decreasing considerably and falling within the standard deviation of the SE group, might suggest that the ELF components of digital mobile telephony signals, play a crucial role in their bio-activity, alone or in conjunction with the RF carrier wave. This is in agreement with the mechanism that we have proposed for the action of EMFs on living organisms, according to which, lower frequency fields are predicted to be more bioactive than higher frequency ones,. According to this mechanism, ELF electric fields of the order of 10^{-3} V/m, are able to disrupt cell function by irregular gating of electrosensitive ion channels on the cell membranes. As shown in Table 4, the ELF components of both GSM 900 and 1800 fields appear to possess sufficient intensity for this, for distances up to 50 cm from the antenna of a mobile phone (or about 50 m from a corresponding base station antenna).

Non-parametric Correlation analysis showed a slightly increased relationship with the RF intensity than with ELF electric field intensity, while Linear Correlation analysis gave an opposite result. A possible conclusion from the Correlation Analysis is that both RF and ELF parameters of the mobile telephony radiations are responsible for the effects, but since non-parametric correlation analysis might be more appropriate because of the non-linearity of our data, perhaps RF is slightly more responsible than ELF. Although the correlation analysis between reproductive capacity and distance from the antenna, gave the same values as between reproductive capacity and RF intensity, distance is only indirectly related to the phenomenon. The effect of the distance depends basically on the fact that the RF and ELF intensities change with the distance. Nevertheless, other possibilities like effect of the radiation wavelength, wave interference, or effect of the differences between near and far field zone of the antenna could not be excluded. These possibilities are investigated and discussed in the following series of experiments together with the nature of the observed bioactivity "windows", (also in Panagopoulos and Margaritis 2010b).

The present set of experiments (a more detailed description can be found in Panagopoulos et al. 2010) showed that, the bioactivity of GSM radiation in regards to short-term exposures is evident for radiation intensities down to 1 $\mu W/cm^2$. This radiation intensity is found at about 1 m distance from a cell-phone or about 100 m distance from a corresponding base station antenna. This radiation intensity is 450 times and 900 times lower than the ICNIRP limits for 900 and 1800 MHz respectively, (ICNIRP, 1998).

It is possible for long-term exposure durations (weeks-months-years) that the effect would be evident at even longer distances/smaller intensities. For this, a safety factor of at least 10 should be introduced in the above value. By introducing a safety factor of 10, the above value becomes 0.1 $\mu W/cm^2$, which should be a reasonable limit for public exposure according to the described findings.

The bioactivity "windows" found in our experiments, could possibly correlate with recent results of another experimental group reporting that GSM radiation caused increased permeability of the blood-brain barrier in rat nerve cells and the strongest effect was produced by the lowest SAR values which correspond to the weakest radiation intensity, (Eberhardt et al. 2008).

Although windows of increased bioactivity of RF radiations have been recorded for many years (Bawin et al. 1975; 1978; Blackman et al, 1980; 1989), there is still no widely accepted explanation for their existence. A novel explanation for the "window" effects is given later on in this chapter.

5. THE DISCOVERED "WINDOW" OF INCREASED BIOACTIVITY IS AN INTENSITY WINDOW

The increased bioactivity "windows" of GSM 900 and 1800 MHz radiations, revealed in the previous experiments, manifesting themselves as a maximum decrease in the reproductive capacity of the insect *Drosophila melanogaster*, were examined in this series of experiments, in order to find out whether they depend on the intensity of the radiation-fields, or to any other possible factor related to the distance from the antenna.

In these experiments, one group of insects (consisting again of ten male and ten female newly eclosed adult flies) were exposed to the GSM 900 or 1800 radiation at 30 cm or 20 cm distances respectively from the antenna of a mobile phone, where the bioactivity "window" appears for each type of radiation and another group was exposed at 8 cm or 5 cm respectively, behind a metal grid, shielding both microwave radiation and the extremely low frequency (ELF) electric and magnetic fields for both types of radiation in a way that radiation and field intensities were roughly equal between the two groups. Then the effect on reproductive capacity was compared between the two groups for each type of radiation.

The average mean values of reproductive capacity (number of F_1 pupae per maternal fly) ± SD from five identical experiments with each kind of radiation are shown in Table 6 and represented in Figure 6.

Table 6. Effect of GSM 900 and 1800 radiation-fields on the Reproductive Capacity of Groups Exposed at "Window" Intensity and Sham Exposed Groups

Groups	Average Mean Number of F_1 Pupae per Maternal Fly ± SD, for GSM 900 MHz, in five replicate experiments	Deviation from SE Group	Average Mean Number of F_1 Pupae per Maternal Fly ± SD, for GSM 1800 MHz, in five replicate experiments	Deviation from SE Group
E1	7.86 ± 0.95	-42.63 %	8.38 ± 0.93	-38.56 %
E2	7.84 ± 0.65	-42.77 %	8.36 ± 0.77	-38.71 %
SE	13.7 ± 0.70		13.64 ± 0.65	

The results show that the reproductive capacity between the two exposed groups did not differ significantly for both types of radiation, ($P > 0.97$ in both cases, meaning that differences between the two exposed groups have more than 97% probability to be due to random variations according to the statistical analysis). In contrast, the reproductive capacity of each exposed group was significantly decreased compared to the sham exposed group as expected, for both types of radiation, ($P < 10^{-5}$ in all cases).

Therefore, since the two exposed groups do not differ significantly between them, although they were exposed at different distances from the antenna but under the same radiation-field intensities, the discovered window of increased bioactivity depends on the intensity of the radiation-field (10 $\mu W/cm^2$, 0.6-0.7 V/m) at 30 cm or 20 cm from the GSM 900 or 1800 mobile phone antenna respectively and not to any other factor that could possibly be related with the certain distances from the antenna. Thus, the increased bioactivity window of digital mobile telephony radiation found in the previous set of experiments is actually an Intensity Window around the value of 10 μW/cm² (in regards to the RF intensity), [or around the values of 0.6-0.7 V/m and 0.10-0.12 mG (in regards to the ELF electric or magnetic field intensities respectively), or to any combination of the three of them]. Within this "window" the bioactivity of mobile telephony radiation becomes even stronger than for intensities higher than 250 μW/cm², (or higher than 13 V/m and 0.6 mG respectively). Under normal conditions and without obstacles between the antenna and the exposed object, the intensity around 10 μW/cm² where the window appears is encountered at a distance of approximately 30 cm from a GSM 900 or 20 cm from a GSM 1800 mobile phone antenna, which corresponds to a distance of about 30 or 20 meters respectively from a corresponding base station antenna, since, as explained, base station antennas emit the same kind of radiation at about 100 times higher power than the corresponding mobile phones, (Panagopoulos et al. 2010; Panagopoulos and Margaritis 2008; Hyland 2000).

Bioactivity of GSM 900 and 1800 Radiation at "Window" Intensity

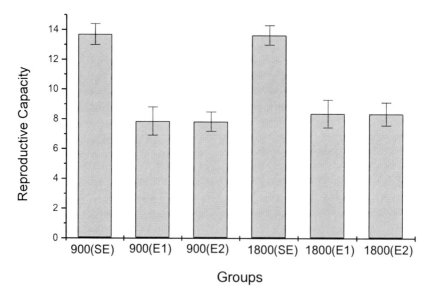

Figure 6. Reproductive Capacity (average mean number of F_1 pupae per maternal insect) ± SD of exposed and sham exposed insect groups to GSM 900 MHz and 1800 MHz radiation at "Window" Intensity (10 μW/cm²). The decrease in reproductive capacity of the exposed groups E1 and E2 for both types of radiation is significant in relation to the sham exposed groups but there is no significant difference between them.

We have thus shown that the discovered window is only indirectly related to the distance from the antenna, therefore it does not seem to be related with the wavelength (or the frequency) of the radiation. This window is directly dependent on the intensity of the radiation/field, no matter on what distance from the antenna this intensity is encountered.

A more detailed description of these experiments can be found in (Panagopoulos and Margaritis 2010b).

6. THE DECREASE IN REPRODUCTIVE CAPACITY IS DUE TO DNA DAMAGE AND CELL DEATH INDUCTION IN THE REPRODUCTIVE CELLS

To determine the ability of GSM 900 and 1800 MHz radiation to act as possible genotoxic factors able to induce DNA damage and/or cell death, at different intensities (or at different distances from the mobile phone antenna), we used TUNEL (Terminal deoxynucleotide transferase dUTP Nick End Labeling) assay.

This is a widely used method for identifying DNA fragmentation and cell death. By the use of this method, a fluorescent substance, fluorescein dUTP, is bound through the action of terminal transferase, onto fragmented genomic DNA which then becomes labelled by characteristic fluorescence. The label incorporated at the fragmented DNA is visualized by fluorescence microscopy (Gavrieli et al, 1992).

The DNA fragmentation test with the use of TUNEL assay, was applied in the ovaries of the exposed and sham-exposed female insects, and especially, in the developing eggs at the stages of early and mid oogenesis, when no programmed cell death takes place, as explained below.

Each *Drosophila* ovary consists of 16 to 20 ovarioles. Each ovariole is an individual egg assembly line, with new egg chambers in the anterior moving toward the posterior as they develop, through 14 successive stages until the mature egg reaches the oviduct. The most anterior region is called the germarium. The most sensitive developmental stages during oogenesis for stress-induced cell death, are region 2 within the germarium and stages 7-8 just before the onset of vitellogenesis (Drummond-Barbosa and Spradling 2001; McCall 2004). Physiological apoptosis (programmed cell death) takes place normally, in the nurse and follicle cells of developing egg chambers during the last stages (11-14) of oogenesis (choriogenesis), (Nezis et al. 2000; 2002; McCall 2004). Additionally, in cases that certain egg chambers do not develop normally, the organism itself destroys them by induction of apoptosis at either one of the two above developmental stages (germarium or stage 7-8) which are called for this reason, "check points". This stress-induced apoptosis, is a vital process in gametogenesis (oogenesis, spermatogenesis) and reproduction, by which the organism prevents the waste of precious nutrients. Previously known external stress factors like chemical stress, heat shock, or poor nutrition, are able to induce cell death during early and mid oogenesis, exclusively in the nurse and the follicle cells of abnormally developing egg chambers, and exclusively at the two check points (Drummond-Barbosa and Spradling 2001; McCall 2004; Nezis et al.2000; Panagopoulos et al. 2007a). No distinction between the two check points was found before, in regards to which one is more sensitive than the other.

Electromagnetic stress from mobile telephony radiations was found in earlier experiments of ours to be extremely bioactive, inducing DNA damage and cell death to a high degree during early and mid oogenesis, not only to the above "check points" (germarium and stages 7-8) but to all the developmental stages of the early and mid oogenesis and moreover to all types of egg chamber cells, i.e. nurse cells, follicle cells and the oocyte (OC) (Panagopoulos et al, 2007a).

Wild-type strain Oregon R *Drosophila melanogaster* flies were cultured according to standard methods and kept in glass vials with standard food like in the previous series of experiments, (Panagopoulos et al, 2004; 2007a; 2010).

In each single experiment of this series, we collected newly eclosed adult flies from the stock early in the afternoon, and separated them into thirteen groups exactly as in the experimental set No 4 (see above), following our standard methodology, (Panagopoulos et al, 2004). We applied TUNEL assay in the ovaries of female insects (for details see, Panagopoulos et al. 2007a; 2010), in order to investigate possible DNA damage at different distances from the mobile phone (in other words, for different intensities of GSM 900 and 1800 radiations).

The total duration of exposure was again, 6 min per day in one dose and the exposures were started on the first day of each experiment. All the 12 exposed groups were simultaneously exposed at the same various distances from the mobile phone as in the experiments No 4 during the 6 min exposure sessions. The exposures took place for five days in each experiment, as previously described, (Panagopoulos et al, 2004). Then there was an additional 6 min exposure in the morning of the sixth day and one hour later, female insects from each group were dissected, and their ovaries were extracted to be prepared for the TUNEL assay as follows:

TUNEL Assay

The ovaries were dissected in Ringer's solution and separated into individual ovarioles from which we excluded all the egg chambers of stages 11-14. As we have already explained, in the egg chambers of stages 11-14 programmed cell death takes place normally in the nurse cells and follicle cells. For this, we kept and treated ovarioles and individual egg chambers from germarium up to stage 10. Samples were fixed in phosphate-buffered saline (PBS) solution containing 4% formaldehyde plus 0.1% Triton X-100 (Sigma Chemical Co., Munich, Germany) for 30min and then rinsed three times and washed twice in PBS for 5 min each. Then samples were incubated with PBS containing 20 μg/ml proteinase K for 10 minutes and washed three times in PBS for 5 min each. In situ detection of fragmented genomic DNA was performed with Boehringer Mannheim kit (Boehringer Mannheim Corp., Indianapolis, IN, USA), containing fluorescein dUTP for 3h at 37°C in the dark. Samples were then washed six times in PBS for 1h and 30 min total duration in the dark and finally mounted in antifading mounting medium (90% glycerol containing 1.4-diazabicyclo (2.2.2) octane (Sigma Chemical Co.) to prevent from fading and viewed under a Nikon Eclipse TE 2000-S fluorescence microscope (Tokyo, Japan).

The samples from different experimental groups were blindly observed under the fluorescence microscope (i.e. the observer did not know the origin of the sample) and the percentage of egg chambers with TUNEL-positive signal was scored in each sample. Statistical analysis was made by single factor Analysis of Variance test.

In Table 7, the summarised data on cell death induction in the gonads of the female insects during early and mid oogenesis from three separate experiments are listed. These data are represented graphically in Figures 7 and 8. The percentages of TUNEL-positive egg chambers in all groups were found to be very close to the corresponding decrease in the reproductive capacity of the same groups (Tables 5, 7, Fig. 4, 5, 7, 8), as in earlier experiments of ours, (Panagopoulos et al.2007a). The maximum percentage of TUNEL-positive egg chambers of exposed animals was found in the ovaries of female insects exposed at 0 and 20 cm distance from the antenna for GSM 1800 MHz (43.39 % and 55.07 %) and at

0 and 30 cm distance correspondingly for GSM 900 MHz (57.72 % and 57.83 %), in agreement with the corresponding maximum decreases in the reproductive capacity (Tables 5, 7, Fig. 4, 5, 7, 8).

Table 7. Effect of GSM 900 and 1800 on ovarian DNA Fragmentation at different Distances from the mobile phone Antenna

Groups - Distance from mob. phone Antenna (cm)	GSM 900 Sum ratio of TUNEL-Positive to Total Number of egg-chambers from germarium to stage 10± SD	Percentage of TUNEL-Positive egg-chambers (%)	Deviation from Sham-Exposed groups (%)	GSM 1800 Sum ratio of TUNEL-Positive to Total Number of egg-chambers from germarium to stage 10± SD	Percentage of TUNEL-Positive egg-chambers (%)	Deviation from Sham-Exposed groups (%)
0	355/615=0.5772 ±0.083	57.72	+50.16	243/560=0.4339 ±0.087	43.39	+35.77
1	267/612=0.4363 ±0.061	43.63	+36.01	146/483=0.3023 ±0.059	30.23	+22.61
10	172/577=0.2981 ±0.052	29.81	+22.24	136/532=0.2556 ±0.054	25.56	+17.94
20	152/564=0.2695 ±0.049	26.95	+19.38	337/612=0.5507 ±0.095	55.07	+47.45
30	336/581=0.5783 ±0.092	57.83	+50.26	78/452=0.1726 ±0.061	17.26	+9.64
40	93/542=0.1716 ±0.053	17.16	+9.59	62/577=0.1075 ±0.056	10.75	+3.13
50	60/556=0.1079 ±0.043	10.79	+3.22	54/511=0.1057 ±0.042	10.57	+2.95
60	51/498=0.1024 ±0.045	10.24	+2.67	57/580=0.0983 ±0.046	9.83	+2.21
70	57/584=0.0976 ±0.041	9.76	+2.19	39/427=0.0913 ±0.033	9.13	+1.51
80	51/563=0.0906 ±0.037	9.06	+1.49	39/485=0.0804 ±0.034	8.04	+0.42
90	50/591=0.0846 ±0.04	8.46	+0.89	41/534=0.0768 ±0.028	7.68	+0.06
100	46/602=0.0764 ±0.035	7.64	+0.07	43/557=0.0772 ±0.035	7.72	+0.1
SE	47/621=0.0757 ±0.038	7.57	0	48/630=0.0762 ±0.034	7.62	0

The effect of cell death induction in the developing eggs of the exposed female insects, just like the corresponding effect on the reproductive capacity, was very intense for distances up to 30 cm from the mobile phone antenna, then diminished considerably for distances longer than 40-50 cm from the mobile phone antenna where the ELF components decrease significantly, but it was still evident for distances up to 100 cm (radiation intensities down to 1 µW/cm^2).

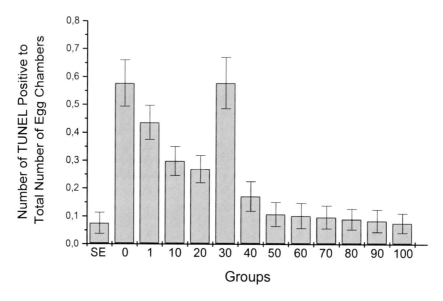

Figure 7. Mean ratio of egg-chambers with fragmented DNA (number of TUNEL- positive to total number of egg-chambers) ± SD, in relation to the Distance from a GSM 900 MHz mobile phone antenna, (cm). The increase in DNA damage and consequent cell death induction is maximum at zero distance and at 30 cm distance from the antenna (intensity "window"), corresponding to RF intensities 378 µW/cm2 and 10 µW/cm2, (Tables 7, 4).

Figure 9a shows an ovariole from a sham exposed (SE) female insect, containing egg chambers from germarium to stage 8, all TUNEL-negative. This was the typical picture in the vast majority of ovarioles and separate egg chambers from female insects of the sham exposed groups. In the SE groups, only few egg chambers (including germaria), (less than 8%), were TUNEL-positive (Table 7, Fig. 7, 8), exclusively at the two check points, a result that is in full agreement with the rate of spontaneously degenerated egg chambers normally observed during *Drosophila* oogenesis, (Nezis et al. 2000; Baum et al. 2005; Panagopoulos et al. 2007a).

Figure 9b shows an ovariole of an exposed female insect (group 50- GSM 900), which is TUNEL-positive only at the two "check points" germarium and stage 7 and TUNEL-negative at all other developmental stages. This was a typical picture of ovarioles of exposed insects from the groups 40 to 90 for GSM 900 and 30 to 80 for GSM 1800.

Figure 9c shows an ovariole of an exposed female insect (group 20- GSM 1800), with a TUNEL-positive signal at all developmental stages from germarium to 8 and in all the cell types of the egg chamber, (nurse cells, follicle cells and the oocyte). This was a usual picture of ovarioles of exposed insects from the groups 0 to 30 for GSM 900 and 0 to 20 for GSM 1800.

Although in most egg-chambers where DNA fragmentation could be observed, the TUNEL-positive signal was most evident in the nurse cells, in many egg chambers of exposed animals and especially in the groups 0 to 30 for GSM 900 and 0 to 20 for GSM 1800 on which the impact of the radiation was maximum, a TUNEL-positive signal was detected in all three kinds of egg chamber cells, (figure 9c).

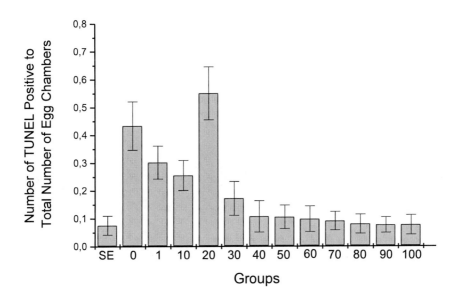

Figure 8. Mean ratio of egg-chambers with fragmented DNA (number of TUNEL- positive to total number of egg-chambers) ± SD, in relation to the Distance from a GSM 1800 MHz mobile phone antenna, (cm). The increase in DNA damage and consequent cell death induction is maximum at zero distance and at 20 cm distance from the antenna (intensity "window"), corresponding to RF intensities 252 µW/cm2 and 11 µW/cm2, (Tables 7, 4).

In the SE groups, random DNA fragmentation was observed exclusively at the two developmental stages named check-points (germarium and stage 7-8) as also observed before, (Panagopoulos et al. 2007a). Similarly, induced DNA fragmentation in the groups 40 to 100 for GSM 900 and 30 to 100 for GSM 1800, (as in fig. 9b), was observed mostly at the two check-points, (data not shown) and only in few cases at the other provitellogenic and vitellogenic stages, 1-6 and 9-10, correspondingly. In contrast, ovarian egg chambers of animals from the exposed groups 0 to 30 for GSM 900 and 0 to 20 for GSM 1800, were found to be TUNEL-positive to a high degree at all developmental stages from germarium to stage 10, (fig. 9c), (data not shown). In all cases (both in the SE and in the exposed groups), the TUNEL-positive signal was observed predominantly and was most intense at the two check points, germarium and stages 7-8, as previously recorded, (Panagopoulos et al. 2007a).

Therefore, we verified that the two check points found by other experimenters (Nezis et al. 2000; 2002; Drummond-Barbosa and Spradling 2001; McCall 2004; Baum et al. 2005) to be the most sensitive developmental stages in regards to other kinds of external stress, are also the most sensitive stages in regards to electromagnetic stress. Moreover, the germarium was found for the first time, to be even more sensitive than the mid-oogenesis check point (stages 7-8) in regard to the electromagnetic stress (Panagopoulos et al. 2007a; 2010).

The effect of GSM radiation-field on DNA damage, and the consequent induced cell death in the ovaries of exposed female insects, diminishes considerably, just as the effect on the reproductive capacity, for distances longer than 40 cm from the mobile phone antenna and disappears for distances longer than 80-90 cm, corresponding to radiation intensities smaller than 1 $\mu W/cm^2$, (Tables 5, 7, Fig. 4, 5, 7, 8). For distances longer than 50 cm where the ELF components decrease significantly and fall within the background of the stray 50 Hz fields, the increase in cell death induction, just as the decrease in reproductive capacity, in regard to the SE groups was very small, falling within the standard deviation of the SE groups, (Tables 5, 7, Fig. 4, 5, 7, 8).

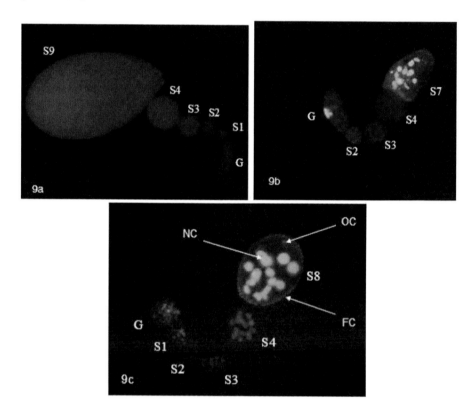

Figure 9. a) Typical TUNEL-negative fluorescent picture of an ovariole of a Sham Exposed female insect, containing egg chambers from germarium to stage 9. b) Ovariole of an exposed insect (group 50-GSM 900) with TUNEL-positive signal only at the two check points, (germarium and stage 7 egg chamber) and TUNEL-negative intermediate stages. c) Ovariole of an exposed female insect (group 20-GSM 1800) with fragmented DNA at all stages from germarium to stage 8), and in all kinds of egg chamber cells, (NC: nurse cells, FC: follicle cells, OC: oocyte).

The statistical analysis (single factor analysis-of-variance test) shows that the probability that cell death induction differs between groups because of random variations, is $P < 10^{-10}$ both for GSM 900 MHz and 1800 MHz exposures. Therefore, the groups differ between them in cell death induction because of the GSM 900/1800 exposures at the different distances-intensities and not due to random variations. The reason that the P value is much smaller in the case of reproductive capacity ($P < 10^{-27}$ in experiments No 4) than in cell death induction ($P < 10^{-10}$), is only that the number of experiments for cell death induction (3) was smaller than the corresponding number of experiments for the effect on reproductive capacity (8).

The increased bioactivity window found in our previous experiments in regard to the effect on the reproductive capacity was also recorded in this set of experiments for the same radiation - field intensity values in regard to DNA damage. We do not know whether this intensity window is related exclusively with the certain organism that we used as experimental animal, or it would appear for other organisms too. More experiments with different experimental animals exposed at different distances from a mobile phone antenna are necessary to answer this question. Nevertheless, since the effect of GSM radiation on DNA damage was observed in all three different kinds of female reproductive cells (nurse cells, follicle cells and the oocyte) and since most cellular functions are identical in both insect and mammalian cells, we consider that it is possible for the above intensity window to exist for other organisms and humans as well.

Our results show that exposure of living organisms to mobile telephony radiation is highly bioactive, able to induce DNA damage and cell death, at intensities higher than few $\mu W/cm^2$ and this bioactivity is still evident for intensities down to $1\mu W/cm^2$, (corresponding to distances up to 100 cm from a mobile phone, or up to about 100 m from a base station antenna). Effects were not observed at intensities lower than 1 $\mu W/cm^2$ in the specific biological system that we studied, in regards to short term exposure periods.

As in earlier experiments of ours (Panagopoulos et al. 2007a), although egg chambers during early and mid oogenesis in *Drosophila* were not reported before to exhibit either stress-induced by other stress factors than EMFs, or physiological degeneration, at other stages except germarium and stages 7-8 (Drummond-Barbosa and Spradling 2001; Nezis et al. 2000; 2002; McCall 2004), mobile telephony radiation was found to induce cell death at all provitellogenic and vitellogenic stages, 1-10 and the germarium. Additionally, again, cell death could be observed in all the cell types of the egg chamber, i.e. not only in nurse cells and follicle cells on which it was already known to be induced by other stress factors than EMFs (McCall 2004; Drummond-Barbosa and Spradling 2001; Nezis et al. 2000; 2002; Cavaliere et al. 1998; Foley and Cooley 1998), but also in the oocyte (fig. 9c).

Thus, electromagnetic stress from mobile telephony radiations was found in our experiments to be much more bioactive than previously known stress factors like poor nutrition, excessive heat or cytotoxic chemicals, inducing cell death to a higher degree not only to the two check points but to all developmental stages of early and mid oogenesis and moreover to all types of egg chamber cells, i.e. nurse cells, follicle cells and the oocyte (OC) (Panagopoulos et al, 2007a).

A possible explanation for these phenomena as given by us before (Panagopoulos et al. 2007a) is that, the electromagnetic stress induced in the ovarian cells by the GSM 900 and 1800 fields, is a new and probably more intense type of external stress, against which ovarian cells do not have adequate defence mechanisms like they do in the case of other kinds of external stress like poor nutrition, heat shock or chemical stress.

The fact that the electromagnetic stress induces DNA fragmentation also in the oocyte (except of the nurse and follicle cells which anyway degenerate physiologically at stages 11-14), shows that the action of the electromagnetic stress is genotoxic and not just a shift of the physiological apoptotic stages in time as someone could possibly think as an alternative explanation. Besides, if it was just a shift of physiological apoptosis towards earlier stages it would seem more likely for the organism to eliminate the defective egg chambers in the existing check points, germarium and stages 7-8, since this is the reason for the existence of these check points.

It is important to remark that DNA fragmentation in the oocyte which undergoes meiosis during the last stages of oogenesis, may result, if not in cell death, in heritable mutations transferred to the next generations after DNA damage and repair (Panagopoulos et al. 2007a). Such a possibility may be even more dangerous than population reduction.

The results of this set of experiments reveal that the large decrease in the reproductive capacity found in our previous experiments after exposure to GSM radiation, is due to elimination of large numbers of egg chambers during early and mid oogenesis, after induction of cell death on their constituent cells, caused by the mobile telephony radiations, at all the different distaces/intensities tested, up to 1m from a mobile phone antenna (or down to 1 $\mu W/cm^2$, radiation intensity).

We do not know if the induced ovarian cell death is apoptosis, i.e. caused by the organism in response to the electromagnetic stress, or necrosis, caused directly by the electromagnetic radiation. This important issue remains under investigation.

7. THE DNA DAMAGE INDUCED BY GSM 900 AND 1800 RADIATION IS ACCOMPANIED BY ACTIN CYTOSKELETON DAMAGE

In this set of experiments (a detailed description can be found in Chavdoula et al. 2010), we showed that GSM radiation, induces disorganization of the actin cytoskeleton in the reproductive cells of exposed female insects during early and mid oogenesis. The disorganization of the actin cytoskeleton is another known aspect of cellular death during both apoptosis and necrosis. For this we applied rhodamine-conjugated phalloidin staining assay, as described below.

We also examined whether follicles with TUNEL-positive signal in their constituent cells had at the same time alterations in their actin cytoskeleton. For this we used double staining with rhodamine-conjugated phalloidin and TUNEL assay at the same samples following the methodology described below.

Rhodamine-Conjugated Phalloidin Staining Assay

Ovaries were dissected in Ringer's solution, fixed in PBS (Invitrogen, USA, 70013-016) containing 4% formaldehyde (Polysciences, Inc., Warrington, PA, 18814) for 20 min, and permeabilized for 35 min in PBS containing 4% formaldehyde plus 0.1% Triton X-100. Individual follicles were then stained for 2 h in PBS containing 1 mg/ml rhodamine-conjugated phalloidin (Invitrogen, USA, R415) (rhoramine is a fluorescent substance that gets attached to the actin cytoskeleton through the binding of phalloidin), washed three times (5 min each) in PBS and finally they were mounted in 90% glycerol containing 1.4-diazabicyclo (2.2.2) octane (Sigma Chemical Co., Germany) to avoid fading (antifading mounting medium).

Double Staining with Rhodamine-Conjugated Phalloidin and TUNEL

Ovaries were dissected in Ringer's solution, fixed in PBS containing 4% formaldehyde for 20 min, and permeabilized for 35 min in PBS containing 4% formaldehyde plus 0.1% Triton X-100. Individual follicles were then stained for 2 h in PBS containing 1 mg/ml rhodamine-conjugated phalloidin and washed three times (5 min each) in PBS. Follicles were then incubated with PBS containing 20μg/ml proteinase K for 10 min. The in situ detection of fragmented genomic DNA was performed with the in situ cell death detection kit (Roche, Mannheim, Germany, 11684795910) by using fluorescein-labeled dUTP for 3 h at 37°C in the dark. Following this procedure, the follicles were washed six times in PBS over the course of 90 min in the dark and mounted in antifading mounting medium.

The simultaneous observation of the two cell death features was accomplished by double action of two different lasers on the samples and observation of the corresponding two types of fluorescence through a Nikon EZ-C1 Confocal Laser Scanning Microscope (CLSM) (Nikon Instruments, Japan).

Figure 10a shows a stage 10 egg chamber from a sham-exposed insect, treated with rhodamine-conjugated phalloidin assay, with normal cytoskeleton morphology. Characteristic features of the actin cytoskeleton in the nurse cells can be observed, like the ring channels (RC) which facilitate the transport of proteins and mRNAs from the nurse cells to the developing oocyte.

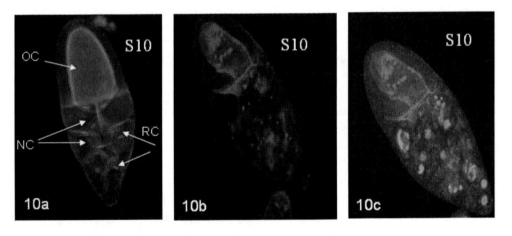

Figure 10. a) A stage 10 egg chamber from a sham exposed insect, treated with rhodamine-conjugated phalloidin assay, with normal cytoskeleton morphology. Characteristic features of the actin cytoskeleton like the ring channels (RC) can be observed. NC: nurse cells, OC: oocyte. b) Stage 10 egg chamber of an exposed insect with disorganized actin cytoskeleton. c) The same stage 10 egg chamber as in figure 10b, treated with both TUNEL (green fluorescence) and rhodamine-conjugated phalloidin (orange fluorescence) assays, revealing that DNA fragmentation and actin cytoskeleton disorganization coexist in the damaged follicles of the exposed insects.

Figure 10b shows a stage 10 egg chamber of an exposed insect with disorganized (damaged) actin cytoskeleton.

Figure 10c shows the same stage 10 egg chamber as in figure 10b, treated with both TUNEL and rhodamine-conjugated phalloidin assays, revealing that DNA fragmentation and actin cytoskeleton disorganization coexist in the damaged follicles of the exposed insects.

This set of experiments demonstrated the simultaneous induction of two cell death features, DNA fragmentation and actin cytoskeleton disorganization in the egg chamber cells, during early and mid oogenesis, when no physiological apoptosis takes place. Both features of cellular death were found to coincide in the damaged follicles, (fig. 10c).

8. THE BIO-EFFECT OF THE GSM RADIATION INCREASES WITH INCREASING DAILY EXPOSURE DURATION

In this set of experiments we examined the bioactivity of different durations of a single, (continuous), daily exposure, ranging from 1 min up to 21 min, to GSM 900 and 1800 radiations. The insects were exposed to each type of radiation at an intensity of about 10 $\mu W/cm^2$, corresponding to a distance of 20 cm or 30 cm from the antenna of a GSM 1800 or a GSM 900 mobile phone handset, respectively. At these distances the bioactivity of mobile telephony radiation was shown to be maximum due to the existence of the intensity window described in the previous sets of experiments.

The duration of exposure to any kind of external stimulus is an important parameter in order to know whether the biological effects related to this stimulus are cumulative or not, i.e. whether there is a difference in exposing an organism for a longer or a shorter time. It is well documented that ionizing radiations have cumulative effects on living organisms as these effects increase with the absorbed dose, i.e. the amount of energy absorbed by the unit mass of tissue (Coggle 1983; Hall and Giaccia 2006). In the case of non-ionizing radiation, and especially the RF-microwave radiation emitted by mobile telephony antennas, only few such studies were performed until the first publication of this set of experiments (Panagopoulos and Margaritis 2010a), in some cases with contradictory results.

The results of this set of experiments showed that the reproductive capacity decreases almost linearly with increasing exposure duration to both GSM 900 and 1800 radiation, suggesting that short-term exposures to these radiations have cumulative effects on living organisms.

A dual band mobile phone was used again, that could be connected to either GSM 900 or 1800 networks simply by changing the SIM ("Subscriber Identity Module") card on the same handset. The highest Specific Absorption Rate (SAR) for the human head, according to the manufacturer, was 0.795 W/Kg, while the corresponding established limit is 2 W/kg, (ICNIRP 1998). The exposure procedure was the same as in No 1 and 3-5 sets of experiments, but a recorder was used as a sound source instead of the experimenter speaking on the mobile phone during the exposures. The mobile phone was operating in speaking mode, with the same recorded voice, reading the same text during the exposures, the sham exposures and the measurements. The sound source/recorder was always at the same position in relation to the mobile phone, the insects or the probe of the field meter. The handset was fully charged before each set of exposures and measurements.

The insects within the glass vials were exposed at 20- and 30-cm distance from the mobile phone antenna to the GSM 1800 and GSM 900 signals, respectively, where the intensity of the modulated radiation (speaking emission) is roughly equal between the two types of radiation, i.e. about 10 $\mu W/cm^2$, and where the bioactivity of this radiation was previously found to reach a maximum, (Panagopoulos and Margaritis 2008; 2009; 2010b; Panagopoulos et al. 2010).

In each experiment with both GSM 900 and 1800 radiations, we separated the insects into six groups: (a) the group exposed to the radiation/field for 1 min (named "E1"), (b) the group exposed for 6 min (named "E2"), (c) the group exposed for 11 min, (named "E3"), (d) the group exposed for 16 min (named "E4"), (e) the group exposed for 21 min, (named "E5") and (f) the sham-exposed group (named "SE").

Each one of the six groups in each experiment consisted of ten female and ten male newly emerged flies, as in the previous sets of experiments. The sham-exposed groups received identical treatment to the exposed ones, except that they were not exposed to any kind of radiation, since the mobile phone was turned off during the sham exposures. The duration of the sham exposures was 21 min; it was already verified that there was no statistically important difference in the reproductive capacity between groups sham-exposed for all the different selected exposure durations from 1 min up to 21 min (data not shown).

During the exposures, the mobile phone was stabilized with its antenna parallel to the axis of the cylindrical glass vials. The insects of the different groups within their glass vials were exposed simultaneously to GSM 900 or 1800, placed along constant intensity sectors of an arc with a 30- or 20-cm radius, respectively, at the center of which the handset was placed. This exposure arrangement was designed in order to have the different groups equally exposed during their common exposure periods, since there are constant changes in the intensity and frequency of the real mobile telephony signals. Then, during each exposure session, the different groups were taken away from the exposure bench one by one, as soon as the exposure duration of each one was completed. After each exposure, the corresponding sham exposure was performed, at the same distance from the mobile phone handset.

We carried out twelve replicate experiments, six with the GSM 900 MHz radiation and six with the GSM 1800 MHz radiation. The results are shown in Table 8 and are represented graphically in Figure 11.

The data show that the reproductive capacity of all the exposed groups is significantly decreased compared with the sham-exposed groups, for both radiation types and for all the exposure periods from 1 min to 21 min.

Moreover, the data show that the reproductive capacity decreases almost proportionally as the exposure duration increases, for both types of mobile telephony radiation.

The average decrease for the six experiments of each series compared with the control (SE) groups, was for GSM 900 MHz, 36.4% for 1 min exposure, 42.5% for 6 min, 49.2% for 11 min, 56.1% for 16 min, 63.0% for 21 min, and, correspondingly, for GSM 1800 MHz, 35.8%, 41.8%, 49.0%, 55.8% and 62.4%, (Table 8; Figure 11). The average decrease was smaller in the GSM 1800 groups than in the GSM 900 groups, for all the selected exposure durations, although differences between the GSM 900 and 1800 corresponding groups were within the standard deviations, (Table 8; Figure 11).

The statistical analysis shows that for both types of radiation, the probability that the reproductive capacity differs between the six groups owing to random variations, is negligible, $P < 10^{-16}$. The corresponding probability between each exposed group and the SE was in all cases $P < 10^{-5}$ for both types of radiation. The corresponding probability between any two exposed groups that differ 5 min in exposure duration (e.g. E1-E2, E2-E3, etc), was in all cases $P < 0.07$ for GSM 900 exposures and $P < 0.08$ for GSM 1800 exposures. Finally, the corresponding probability between groups that differ 10 min in exposure duration between them (e.g. E1-E3, E2-E4 etc), was in all cases $P < 10^{-2}$.

We did not detect any temperature increases, within the glass vials during the exposures, for all the different exposure durations tested.

Table 8. Effect of different Exposure Durations of GSM 900 and 1800 radiation on the Reproductive Capacity of Drosophila melanogaster

Type of Radiation	Groups (daily Exposure Duration)	Average Mean Number of F_1 Pupae per Maternal Fly ± SD, in six identical experiments	Deviation from Sham Exposed (SE) groups
GSM 900	SE (0 min)	13.10 ± 0.95	
	E1 (1 min)	8.33 ± 0.71	-36.4 %
	E2 (6 min)	7.53 ± 0.60	-42.5 %
	E3 (11 min)	6.65 ± 0.63	-49.2 %
	E4 (16 min)	5.75 ± 0.62	-56.1 %
	E5 (21 min)	4.85 ± 0.69	-63.0 %
GSM 1800	SE (0 min)	13.05 ± 0.96	
	E1 (1 min)	8.38 ± 0.72	-35.8 %
	E2 (6 min)	7.60 ± 0.66	-41.8 %
	E3 (11 min)	6.65 ± 0.61	-49.0 %
	E4 (16 min)	5.77 ± 0.73	-55.8 %
	E5 (21 min)	4.90 ± 0.67	-62.4 %

The statistical analysis clearly shows that the exposed *Drosophila* groups differ in offspring production between themselves and compared with the sham-exposed groups, and this difference is not due to random variations but due to the effect of the GSM fields. The reason why differences between groups differing only 5 min in daily exposure duration were not as strong ($P < 0.07$ for GSM 900 and $P < 0.08$ for GSM 1800) as between groups differing 10 min or more in exposure duration ($P < 10^{-2}$), is only that a 5-min difference in daily exposure duration was not enough to show a large difference in reproductive capacity, since all the exposures (even the shortest of 1 min daily) produced a significant effect and all the exposed groups were significantly different from the sham-exposed groups ($P < 10^{-5}$). When the difference in exposure duration is increased to 10 min or more, then the difference in reproductive capacity between exposed groups becomes highly significant.

The effect of both types of digital mobile telephony radiation on the reproductive capacity seems to increase almost linearly with increasing exposure duration from 1 to 21 min, suggesting that short-term exposures to this radiation have cumulative effects on living organisms.

We do not know whether longer short-term exposures than 21 min would result in an even higher decrease of reproductive capacity, or whether the effect would saturate after a certain exposure duration. This is left for future experiments. Our present results represent a first indication that this radiation can have cumulative effects.

GSM 900/1800 Exposure Duration effect on Reproductive Capacity

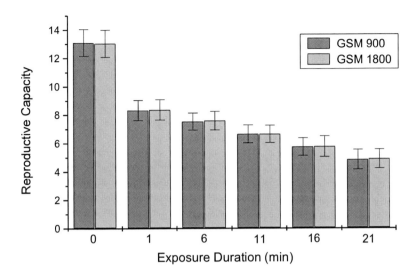

Figure 11. Reproductive Capacity (average mean number of F_1 pupae per maternal insect) ± SD of insect groups exposed to GSM 900 and GSM 1800 radiations for different daily exposure durations (1, 6, 11, 16, and 21 min) and of sham exposed groups (0 min).

It is also unknown whether long-term exposures (for weeks, months, years) of people or animals residing in areas exposed to radiation by base-station antennas also have cumulative effects, but this is possible as the nature of radiation from base stations does not differ from that of mobile phones (Panagopoulos et al. 2010; Hillebrand 2002; Clark 2001; Hamnerius and Uddmar 2000; Tisal 1998). Additionally, the radiation intensity used in the present experiments (~10 $\mu W/cm^2$) is usually encountered at about 20-30 m distance from mobile telephony base-station antennas where people reside or work and therefore may be exposed for up to 24 h per day.

Since we did not detect any temperature increases even during the longest exposures of 21 min, the recorded effect is considered as non-thermal. In previous experiments we exposed the same experimental animal to the near field of a mobile phone antenna (at 0 or 1 cm distances) for exposure periods up to 6 min and the recorded effects were non-thermal (Panagopoulos et al. 2004; 2007a; 2007b). The effects were also non-thermal with six min exposures at different distances in the far field (Panagopoulos et al. 2010). In the present experiments we exposed the insects in the far field of the mobile phone antenna, at 30 and 20 cm, respectively, for 900 and 1800 MHz, for exposure periods up to 21 min and the recorded effects were again non-thermal.

Although both types of radiation considerably affect reproduction, GSM 900 is found again to be slightly more bioactive than GSM 1800, even under equal radiation intensities and for all the exposure durations tested, although the differences in bioactivity between the two types of radiation were within the standard deviation (Table 8, Figure 11).

Our experiments with insects showed that the effect of GSM radiation is cumulative (increases with exposure duration) at least for short-term exposures. According to another study, constant exposure of mice for about 6 months to low-intensity (0.168-1.053 $\mu W/cm^2$) RF radiation from an antenna park had also cumulative effects, and resulted in progressively lower number of newborns from the first to the fifth pregnancy and finally in permanent infertility of the parent animals, (Magras and Xenos 1997).

Our results are also in agreement with other studies that have found a connection between the duration of exposure to mobile telephony radiations and increased health risks (Agarwal et al. 2008; Wdowiak et al. 2007; Salama et al. 2004; Yadav and Sharma 2008), although some contradictory findings are also reported (Zeni et al. 2008; Zareen et al. 2009). Moreover, our present results on insect reproduction are in agreement with certain results on human reproductive capacity (Agarwal et al. 2008; Wdowiak et al. 2007). In both cases, increased exposure duration to mobile telephony radiation induced increased infertility to both insects and humans.

Thus, according to the results of this set of experiments, RF radiations used in modern telecommunication systems seem to have a cumulative biological action and, for this reason, living organisms should be exposed for as short a time as possible to these radiations. Users should be informed to make cautious use of mobile phones and try to diminish the length and frequency of their phone calls.

BASIC EXPERIMENTAL CONCLUSIONS

All the above presented experimental results acquired during the last 12 years have led to the following basic conclusions:

1) GSM 900 and 1800 MHz mobile telephony radiation is found to reduce insect reproduction by up to 60%. The insects were exposed for 6 min daily during the first 5 days of their adult lives. Both males and females were found to be affected.
2) GSM 900 MHz radiation is found to be more bioactive than GSM 1800 MHz under actual conditions mainly due to the fact that, GSM 900 is emitted at double the output power than GSM 1800 (and GSM 1900). GSM 900 is slightly more bioactive than 1800 even under the same intensity.
3) The reduction of insect reproductive capacity is due to DNA damage, actin cytoskeleton damage and cell death induction in the reproductive cells (gonads).
4) The effect in regard to short-term exposures is evident for radiation intensities down to 1 $\mu W/cm^2$. This radiation intensity is found at about 1 m distance from a mobile phone or about 100 m distance from a corresponding base station antenna. This radiation intensity is 450 and 900 times lower than the current ICNIRP limits for 900 and 1800 MHz respectively (ICNIRP 1998). It is possible for long-term exposure durations, that the effect would be evident at even longer distances/smaller intensities. For this, a safety factor should be necessarily introduced in the above value. By introducing a safety factor of 10, the above value becomes 0.1 $\mu W/cm^2$, which is the limit that reasonably results from our experiments and coincides with the limit proposed by the BioInitiative Report (2007).
5) The effect is strongest for intensities higher than 200 $\mu W/cm^2$ (0-1 cm distance from the cell phone) and within a "window" around 10 $\mu W/cm^2$ where the effect

becomes even stronger. This intensity value of 10 μW/cm^2 corresponds to a distance of 20-30 cm from a mobile phone handset or to 20-30 m from a base station antenna.

6) Electromagnetic stress seems to be even more bioactive than other previously tested stress factors like poor nutrition, heat, or chemical stress, inducing DNA damage to a higher degree on insect reproductive cells.
7) The effect increases with increasing daily exposure duration in regard to short-term exposures.
8) The effect is non-thermal- there are no temperature increases during the exposures.
9) The effect at cellular level seems to be due to irregular gating of ion channels on the cell membranes caused by the EMFs, leading to disruption of cell's electrochemical balance and function. This mechanism (presented below) is non-thermal.
10) Although we cannot simply extrapolate the above results from insects to humans, similar effects on humans cannot be excluded. On the contrary, they are possible 1) because insects are in general much more resistant to radiation than mammals and 2) because the presented findings are in intriguing agreement with results of other experimenters, reporting DNA damage on mammalian cells or mammalian (including human) infertility (see below).
11) Reported observations during the last years regarding reduction of bird and insect populations (especially bees) can be explained by decrease in their reproductive capacity as described in our experiments.
12) Symptoms referred to as "microwave syndrome" (headaches, sleep disturbances, fatigue etc.), among people residing around base station antennas, can possibly be explained by cellular stress induction on brain cells or even cell death induction on a number of brain cells.

A DISTINCT SIMILARITY OF EXPERIMENTAL RESULTS

Our experiments published in 2007 (Panagopoulos et al. 2007a) were the first to show that GSM 900 and 1800 radiations emitted by commercially available mobile phones, induce DNA fragmentation and consequent cell death in reproductive cells after *in vivo* exposure of newly eclosed insects. Other experiments before, had indicated DNA damage (Diem et al. 2005) or cell damage (Salford et al. 2003; Markova et al. 2005; Aitken et al. 2005) induced by GSM "test" or simulated signals. Although these signals are significantly different from the real GSM signals that we used, an unquestionable similarity with our results is noticed. A similarity is also noticed with even older results reporting DNA damage in rat brain cells after exposure to microwaves from analog (1st generation) mobile phones (Lai and Singh 1995; 1996; 1997).

Recent results of other experimenters regarding mammalian (including human) infertility, (Gul et al. 2009; Agarwal et al. 2008; Wdowiak et al. 2007; Magras and Xenos 1997) or chicken embryonic mortality (Batellier et al. 2008; Grigor'ev IuG. 2003), especially those regarding DNA damage or oxidative stress on mammalian-human reproductive cells,

(De Iouliis et al. 2009; Mailankot et al. 2009; Yan et al. 2007; Agarwal et al. 2009), exhibit an even more distinct similarity with our results, than the above mentioned (in the previous paragraph) older experiments.

More specifically, GSM radiation from mobile phones decreased the number of follicles in the ovaries of newborn female rats, exposed during their intra-uterine life, (Gul et al. 2009). These results are very similar with the elimination of follicles in the ovaries of female insects after exposure to GSM radiation shown in our experiments. Other recent experiments found that GSM mobile phone radiation induced DNA damage and mitochondrial generation of reactive oxygen species (ROS) in human spermatozoa *in vitro* (De Iuliis et al. 2009; Agarwal et al. 2009). Induction of oxidative stress and reduction of sperm motility in rats are reported in other recent experiments after *in vivo* exposure to real mobile phone 900/1800 radiation, (Mailankot et al. 2009). Sperm cell death in rats after *in vivo* exposure to mobile phone radiation, is also a very similar recent result of other experimenters (Yan et al. 2007).

There is also an agreement of our results with older results of other experimenters reporting DNA damage in cell types other than reproductive, assessed by different methods than ours, after *in vivo* or *in vitro* exposure to GSM radiation, (Sokolovic et al. 2008; Diem et al., 2005; Markova et al., 2005; Salford et al., 2003; Lai and Singh 1995; 1996).

The distinct similarity of the above described results with ours, although in different animals and some of them in different cell types, makes unlikely the possibility that these findings can be wrong. Therefore, it seems that it is an unquestionable fact that microwaves emitted by modern mobile telecommunications, may damage DNA and induce cell death, especially on reproductive cells (gonads) but not only. Additionally, DNA damage and cell death in reproductive cells may explain the recently reported population declinations of bees and birds (Stindl and Stindl 2010; Bacandritsos et al. 2010; van Engelsdorp et al. 2008; Everaert and Bauwens 2007; Balmori 2005).

DNA damage in somatic cells may result in cancer induction. Experimental findings reporting DNA damage in somatic cells seem to be in agreement with recent reports about brain tumor induction among mobile phone users, especially among those with use for more than 10 years (Khurana et al. 2009; Hardell et al. 2007; 2009).

BIOPHYSICAL MECHANISM FOR EMFs BIO-EFFECTS: THE ION FORCED-VIBRATION THEORY

The effects of EMFs at cellular level can be explained by irregular gating of electrosensitive ion channels on the cell membranes, according to the Ion Forced-Vibration Theory that we have proposed (Panagopoulos et al. 2000; 2002; Panagopoulos and Margaritis 2003). This irregular gating of ion channels may lead, non-thermally, to disruption of the cell's electrochemical balance and function, as described below.

According to this theory (Panagopoulos et al. 2000; 2002; Panagopoulos and Margaritis 2003), which is considered until now as the most valid one from all the proposed theories, (Creasey and Goldberg, 2001), even very weak ELF electric fields of the order of 10^{-3} V/m, are theoretically able to change the intracellular ionic concentrations and thus, disrupt cell function. Since RF-microwave radiations and especially those used in modern mobile telecommunications are transmitted within ELF pulses, or include ELF modulating signals of intensities thousands of times higher than 10^{-3} V/m, this theory can be applied for the explanation of their bio-effects.

The basic idea relates to the fact that any external oscillating electric or magnetic field, will induce a forced-vibration on the free ions that exist in large concentrations inside and outside of all living cells in biological tissue. When the amplitude of this forced-vibration exceeds some critical value, the electrostatic force exerted by the oscillating ions' charge on the electric sensors of the voltage-gated membrane ion channels, can irregularly gate these channels, resulting in changes to the free ions' intracellular concentrations.

These free ions play a key role in all cellular functions, initiating or accompanying all cellular biochemical processes.

Let us consider an external oscillating electric field (or the electric component of an electromagnetic wave) of intensity E, acting on a free ion in the vicinity of a cell membrane.

The forced-vibration movement of each free ion due to the external oscillating field, is described by the equation,

$$m_i \frac{d^2x}{dt^2} + \lambda \frac{dx}{dt} + m_i \omega_o^2 x = E_o z q_e \sin\omega t \qquad [1]$$

for the case of an external harmonically oscillating electric field : $E = E_o \sin\omega t$ with circular frequency: $\omega = 2\pi\nu$, (ν, the frequency in Hz), where: z is the ion's valence, $q_e = 1.6\times10^{-19}$ C, the elementary charge, $F_2 = -m_i\omega_o^2 x$, a restoration force proportional to the displacement distance x of the free ion, m_i the ion's mass and $\omega_o = 2\pi\nu_o$, with ν_o the ion's oscillation self-frequency if the ion was left free after its displacement x. In our case, this restoration force is found to be very small compared to the other forces and thus does not play an important role. $F_3 = -\lambda u$ is the damping force, where $u = \frac{dx}{dt}$, the ion's velocity and λ, the attenuation coefficient for the ion's movement, which for the cytoplasm or the extracellular medium is calculated to be $\lambda \cong 10^{-12}$ Kg/sec, while for ions moving inside channel proteins, is calculated to have a value: $\lambda \cong 6.4\times10^{-12}$ Kg/sec, (for Na^+ ions moving through open Na^+ channels), (Panagopoulos et al. 2000).

Assuming that the ions' self frequencies coincide with the frequencies of the cytosolic free ions' spontaneous oscillations observed as membrane potential spontaneous oscillations in many different types of cells with values smaller than 1 Hz and assuming that the ion's maximum oscillation velocity has a value of 0.25 m/s, as calculated for the movement of sodium ions through open sodium channels using patch-clamp conductivity data (Panagopoulos et al. 2000), it comes that the general solution of equation [1], is:

$$x = \frac{E_o z q_e}{\lambda\omega} \cos\omega t - \frac{E_o z q_e}{\lambda\omega} \qquad [2]$$

Since the second term of the second member of eq.[2] is constant, the vibrational movement is described by the equation:

$$x = \frac{E_o z q_e}{\lambda \omega} \cos \omega t \qquad [3]$$

Equation [2] declares that, at the moment when the external field is applied and at the moment when it is interrupted, the displacement of the ion becomes twofold the amplitude of the forced-vibration.

Equation [3] shows that the forced - vibration is in phase with the external force. The amplitude of the free ion forced-vibration is,

$$A = \frac{E_o z q_e}{\lambda \omega} \qquad [4]$$

Thus, the amplitude is proportional to the intensity, and inversely proportional to the frequency of the external oscillating field.

Once this amplitude exceeds some critical value, the coherent forces that the ions exert on the voltage sensors of voltage-gated membrane channels can trigger the irregular opening or closing of these channels, thus disrupting the cell's electrochemical balance and function.

The oscillating ions represent a periodical displacement of electric charge, able to exert forces on every fixed charge of the membrane, like the charges on the voltage sensors of voltage - gated ion channels.

Voltage-gated ion channels, are leak cation channels. The state of these channels, (open/closed), is determined by electrostatic interaction between the channels' voltage sensors, and the transmembrane voltage (which is called also "membrane potential"). They interconvert between open and closed state, when the electrostatic force, exerted by transmembrane voltage changes on the electric charges of their voltage sensors, transcends some critical value. The voltage sensors of these channels, are four symmetrically arranged, transmembrane, positively charged helical domains, each one designated S4 (Noda et al. 1986; Stuhmer et al. 1989).

It is known that changes of about 30 mV in the transmembrane voltage, are able to gate these electrosensitive channels by exerting the necessary electrostatic force on the fixed charges of the S4 helices (Bezanilla et al. 1982; Liman et al. 1991).

We have shown that a single ion's displacement $\partial r \sim 10^{-12}$ m, in the vicinity of S4, can exert an electrostatic force on each S4, equal to that exerted by a change of 30 mV, in the transmembrane voltage (Panagopoulos et al. 2000):

The intensity of the transmembrane electric field is

$$E_m = \frac{\Delta \Psi}{s} \qquad [5]$$

where, $\Delta \Psi$ is the transmembrane voltage and s the membrane's width.

Additionally,

$$E_m = \frac{F}{q} \qquad [6]$$

where F in this case, is the force acting on an S4 domain and q is the effective charge on each S4, which is estimated to have a value,

$$q \cong 1.7 \, q_e \qquad [7]$$

according to the available data (Liman et al. 1991). From equations [5], [6], we get:

$$F = \frac{\Delta\Psi}{s} q \Rightarrow \partial F = \partial \Delta\Psi \frac{q}{s} \qquad [8]$$

(where $\partial \Delta\Psi$ is the change in the transmembrane voltage, that is necessary to gate the channel). For $\partial \Delta\Psi = 30$ mV, $s = 10^{-8}$ m and substituting q from [7], equation [8] gives: $\partial F = 8.16 \times 10^{-13}$ N.

This is the force, on the voltage sensor of a voltage-gated channel, required normally, to interconvert the channel between closed and open state.

The force acting on the effective charge of an S4 domain, via an oscillating, free z-valence cation, is: $F = \dfrac{1}{4\pi\varepsilon\varepsilon_o} \cdot \dfrac{q \cdot zq_e}{r^2} \Rightarrow$

$$\partial F = -2 \cdot \frac{1}{4\pi\varepsilon\varepsilon_o} \cdot \frac{q \cdot zq_e}{r^3} \partial r \Rightarrow \text{(ignoring the minus sign)},$$

$$\partial r = \frac{2\pi\varepsilon\varepsilon_o \partial F \cdot r^3}{q \cdot zq_e}. \qquad [9]$$

This is the minimum displacement of a single, z-valence cation, in the vicinity of S4, able to generate the necessary force ∂F, to gate the channel. Where: r, is the distance between a free ion with charge zq_e and the effective charge q on each S4 domain, which can be conservatively taken as 1 nm (Panagopoulos et al. 2000). $\varepsilon_o = 8.854 \times 10^{-12}$ N^{-1}·m^{-2}·C^2, is the dielectric constant of vacuum.. The relative dielectric constant ε can have a value 80 for a water-like medium, (cytoplasm, or extracellular space), or a value as low as 4, for ions moving inside channel-proteins (Panagopoulos et al. 2000; Honig et al. 1986).

The concentration of free ions on both sides of mammalian cell membranes, is about 1 ion per nm^3 (Alberts et al. 1994). Let us conservatively calculate ∂r for one single-valence cation, interacting with an S4 domain. If two or more single-valence cations interact, (in phase), with an S4 domain, from 1nm distance, ∂r decreases proportionally. For ions moving inside channel-proteins, we assume, that they move in single file, (Palmer 1986; Panagopoulos et al. 2000).

From equation [9] and for $\partial F = 8.16 \times 10^{-13}$ N, we get:

$$\partial r \cong 0.8 \times 10^{-10} \text{ m, (for } \varepsilon = 80)$$

and:

$$\partial r \cong 4\times 10^{-12} \text{ m, (for } \varepsilon = 4) \qquad [10]$$

We can see that, a single cation's displacement of only few picometers from its initial position, is able to interconvert voltage-gated channels, between open and closed states, (for cations moving or bound within channels).

Therefore, any external field, which can induce a forced-vibration on the ions, with an amplitude $A \geq 4 \times 10^{-12}$ m, is able to disrupt the cell's function. Substituting A from eq. [4], in the last condition, it comes that, a bioactive, external, oscillating electric field, of intensity amplitude E_o and circular frequency ω, which induces a forced-vibration on every single-valence ion, ($z=1$), must satisfy the condition:

$$\frac{E_o q_e}{\lambda \omega} \geq 4 \times 10^{-12} \text{ m}. \qquad [11]$$

Since we adopted a value for ∂r, ($\cong 4\times 10^{-12}$ m), valid for cations within channels, (where $\varepsilon = 4$), we shall use the corresponding value for λ, calculated also for cations moving within channels, (Panagopoulos et al. 2000), $\lambda \cong 6.4\times 10^{-12}$ Kg/sec.

Thereby, the last condition becomes:

$$E_o \geq \omega \cdot 1.6 \times 10^{-4} \qquad [12]$$

or

$$E_o \geq \nu \times 10^{-3} \qquad [13]$$

(ν in Hz, E_o in V/m)

Moreover, we have shown that in the most bioactive case of pulsed fields, and for two double valence cations (i.e. Ca^{+2}) interacting simultaneously with the channel sensor, the condition for irregular gating of the channel becomes:

$$E_o \geq \nu \times 0.625 \times 10^{-4} \qquad [14]$$

(ν in Hz, E_o in V/m). Whenever the condition [14] is satisfied, the external field E can irregularly gate cation channels.

Condition [14] declares that, external ELF electric fields with intensities smaller than tenths of a mV/m should theoretically be able to disrupt cell function by irregular gating of ion channels.

According to this mechanism, lower frequency fields are the most bioactive ones and additionally pulsed fields are shown to be more bioactive than continuous, (uninterrupted), ones because of the constant term in eq. [2] which doubles the displacement of the oscillating ions at the onset and at the end of every pulse, (Panagopoulos et al., 2002).

Thereby, the ELF pulses of the mobile telephony signals are certainly within the criteria of this theory and thus able to produce the reported effects on living organisms.

Microwave radiations are always pulsed or modulated on ELF frequencies like in mobile telephony signals in order to be able to carry and transmit information. Therefore the Ion Forced-Vibration theory described above is applicable for the biological effects of the Radio Frequency (RF)-microwave radiations.

The Thermal Noise Problem

Free ions move anyway because of thermal activity, with kinetic energies much larger normally, than the ones acquired due to the action of an external electromagnetic field at intensities encountered in the human environment. In such a case, it has been claimed (Adair 1991) that this thermal motion masks the motion induced by the external field, making this motion unable to produce any biological effect.

But as we have explained (Panagopoulos et al. 2000; 2002), thermal motion is a random motion, in every possible direction, different for every single ion, causing no displacement of the ionic "cloud" and for this it does not play any important role in the gating of channels, or in the passing of the ions through them. On the contrary, forced-vibration is a coherent motion of all the ions together in phase. The thermal motion of each ion and moreover the thermal motions of different ions, result in mutually extinguishing forces on the voltage sensor of an electro-sensitive ion channel, while the coherent-parallel motion of the forced-vibration results in additive forces on the voltage sensor.

Therefore, if two or more cations interact, (in phase), with an S4 domain, from 1nm distance, then the minimum displacement of the ions necessary to gate the channel, ∂r in eq. [9], decreases proportionally. The concentration of free ions on both sides of mammalian cell membranes, is about one ion per nm^3, (Alberts et al. 1994) and for this, we have initially calculated ∂r for one cation, interacting with an S4 domain, although it is very likely that several ions interact simultaneously each moment with an S4 domain from a distance of the order of 1nm. This counts also for the ions moving already within a channel, since it is known that, although they pass through the narrowest part of the channel in single file, (Miller 2000; Palmer 1986; Panagopoulos et al. 2002), several ions fill the pore each moment as they pass sequentially, and several ion-binding sites (three in potassium channels) lie in single file through the pore, close enough that the ions electrostatically repel each other, (Miller 2000).

In the mildest case, if we consider a single ion interacting with an S4 domain, this ion moving with a drift velocity, u = 0.25 m/s, (Panagopoulos et al. 2000), it needs a time interval $\delta t = \dfrac{\partial r}{u} \cong 1.6 \times 10^{-11}$ s, in order to be displaced at the necessary distance $\partial r = 4 \times 10^{-12}$ m. During this time interval δt, this ion will be also displaced because of thermal motion, at a total distance X_{kT}, ranging from 1.6 to 4×10^{-10} m, according to the equation: $X_{kT} = \sqrt{\dfrac{2kT\delta t}{\lambda}}$, for human body temperature, 37°C or T=310 °K. (X_{kT} in m, δt in s, λ in kg/s, $k = 1.381 \times 10^{-23}$ J·K^{-1} the Boltzmann constant), (Panagopoulos et al. 2002).

The ions' mean free path in the aqueous solutions around the membrane is about 10^{-10} m, (Chianbrera et al. 1994), and it is certainly smaller within the channels, (the diameter of a potassium ion is about 2.66×10^{-10} m and the diameter of the narrowest part of a potassium channel is about 3×10^{-10} m, thereby the mean free path of a potassium ion within the channel has to be of the order of 10^{-11} m), (Panagopoulos et al. 2002; Miller 2000).

Therefore the ion within the above time interval δt, will run because of its thermal activity, several mean free paths, each one in a different direction, exerting mutually extinguishing opposing forces on the channel's sensors, while at the same time the ion's displacement because of the external field is in a certain direction, exerting on each S4 domain a force of constant direction.

In the most realistic case, if we consider several ions interacting simultaneously with an S4 domain, then the effect of the external field is multiplied by the number of ions, whereas the effect of their random thermal motions becomes even more negligible.

Thus, the claims that thermal motion masks the displacements of the free ions, caused by an external electric field, if these displacements are smaller than those caused by thermal motion (Adair 1991), are not valid according to the above analysis.

A Novel Possible Explanation of the Bioactivity "Windows"

According to the Ion Forced-Vibration theory, the action of external EMFs on cells is dependent on the irregular gating of membrane electrosensitive ion channels whenever an electric force on the channel sensors exceeds the force exerted on them by a change in the membrane potential of about 30 mV which is necessary to gate the channel normally. If in some kind of cells there is an upper limit for this value of membrane potential change, then the channel would be gated whenever the force exerted on its sensors is within this "window".

For example, the intensity window that we have recorded, in terms of the ELF electric field intensity, is around 0.6-0.7 V/m, (Panagopoulos et al. 2010; Panagopoulos and Margaritis 2010b). Let us assume that it ranges from 0.5 to 1 V/m. According to our theory, these limits correspond to a single-valence, single ion displacement between $\partial r_1 = 1.3 \times 10^{-11}$ m and $\partial r_2 = 2.6 \times 10^{-11}$ m, in the vicinity of the channel's sensors, equal to the amplitude of the induced forced-vibration in each case, according to the equation [4], $\partial r = A = \dfrac{E_o z q_e}{\lambda \omega}$,

where: E_o the amplitude of the external oscillating electric field which is equal to $E\sqrt{2}$ where E the measured (root mean square) value of electric field intensity, z the ion's valence (for example $z = 1$ for K^+ ions), q_e the unit charge ($= 1.6 \times 10^{-19}$ C), $\lambda \cong 6.4 \times 10^{-12}$ Kg/s the attenuation coefficient for the ion movement within a cation channel, $\omega = 2\pi \nu$ (ν the frequency of the external oscillating field, in our case let us accept, $\nu = 217$ Hz the pulse repetition frequency of the GSM signals).

These displacements ∂r_1 and ∂r_2 would exert on each channel's sensor (S4 domain) corresponding forces $\partial F_1 = 2.5 \times 10^{-12}$ N and $\partial F_2 = 5 \times 10^{-12}$ N according to Equation [9], $\partial r = -\dfrac{2\pi \varepsilon \varepsilon_o \partial F \cdot r^3}{q \cdot z q_e}$, where $\varepsilon = 4$, the relative dielectric constant in the internal of a channel-protein, $\varepsilon_o = 8.854 \times 10^{-12}$ N$^{-1} \cdot$m$^{-2} \cdot$C^2, $r \cong 10^{-9}$ m, and $q = 1.7 q_e$, (Panagopoulos et al. 2002; 2000).

A force between 2.5 and 5×10^{-12} N on the channel's sensor, in turn, corresponds according to [8], $\partial F = \partial \Delta \Psi \frac{q}{s}$, to a change $\partial \Delta \Psi$ in the transmembrane voltage between 90 and 180 mV, (for $q = 1.7 q_e$ and $s \cong 10^{-8}$ m the membrane's width).

Thus we have shown that the intensity window found in our recent experiments corresponds to a gating voltage change between 90 and 180 mV in the membrane potential.

Channel gating is usually studied on nerve cells and in this kind of cells possibly no upper limit exists for the gating voltage change, but the possibility of an upper limit (like the value of 180 mV that we found in our example), cannot be excluded for other kinds of cells which have not been studied yet in terms of their channel voltage gating. This hypothesis of ours for the explanation of the existence of bioactivity "windows" was reported recently (Panagopoulos and Margaritis 2010b) for the first time. The given numerical example is just an indication that the bioactivity windows reported since many years in bioelectromagnetic experiments but not explained so far, can possibly be explained according to the Ion Forced-Vibration theory.

BIOCHEMICAL MECHANISMS FOR DNA DAMAGE

Since microwaves are non-ionizing radiations (i.e. do not have the ability to detach electrons from molecules or break chemical bonds) it is unlikely that they can directly break DNA chains. It is possible though, for the ELF pulses of the low frequency modulation signals that co-exist with the microwave carrier, to alter the intracellular ionic concentrations by irregular gating of electrosensitive cation channels on the cell membranes, according to the above described mechanism. This in turn, may initiate the following possible processes:

1. Irregular Release of Hydrolytic Enzymes

It is known that alteration of intracellular ionic concentrations, especially Ca^{+2} may initiate cell death induction through apoptosis or necrosis (Santini et al. 2005). A common event preceding both apoptosis and necrosis, is the increase of mitochondrial calcium ion concentration released by endoplasmic reticulum (Armstrong 2006). The mitochondrial concentration of calcium ions can be increased by irregular uptake due to direct action of the external EMF on calcium channels of the mitochondrial membrane, or indirectly due to increased calcium release in the cytoplasm by endoplasmic reticulum membrane or by plasma membrane, according to the biophysical mechanism described above. These processes may possibly lead to the release of specific hydrolytic enzymes (like DNases) by the cytoplasmic organelles called lysosomes (Goldsworthy 2007). The release of such enzymes may lead in turn to DNA fragmentation. Release of DNases or other hydrolytic enzymes from the lysosomes is mediated by alterations in the intracellular calcium concentrations (Santini et al. 2005; Armstrong 2006; Goldsworthy 2007). DNA fragmentation and consequent cell death induction, as it is shown in the above presented experiments of ours, is the reason for the decrease in the reproductive capacity of insects caused by mobile telephony radiations. Since an external oscillating electromagnetic field can change the intracellular ionic concentrations by irregular gating of ion channels on cell membranes, this may lead to DNA fragmentation and cell death through the irregular release of hydrolytic enzymes.

2. Free Radical Action

Another way for indirect DNA damage by EMFs is through the action of free radicals. It is well known that ionizing radiations can detach electrons from different molecules or break molecular bonds, and form free radicals which in turn may react chemically with different biomolecules including nucleic acids (Coggle 1983; Hall and Giaccia 2006). There is recent evidence of excessive free radical formation after RF-microwave exposures (Phillips et al. 2009; De Iouliis et al. 2009). Since the most abundant molecule in biological cells is that of water (H_2O), microwave radiation can possibly lead to the formation of water free radicals like OH•, O_2H•, H•. These molecules are extremely reactive, having a strong trend to react chemically with different biomolecules including DNA, because of an unpaired electron that they comprise, (symbolized by •). This unpaired-single electron tends to be paired and thus free radicals tend to react with other molecules in order to give, take, or contribute one electron and become stable.

The above mentioned water free radicals, except for their possible direct formation by RF-microwave exposure, can be formatted by hydrogen peroxide (H_2O_2), a product of oxidative respiration in the mitochondria, which can be converted by electromagnetic radiation (EMR) into hydroxyl free radical via the Fenton reaction, a reaction catalyzed by iron within the cells:

$$H_2O_2 + (EMR) \xrightarrow{Fe} OH\bullet + OH\bullet \qquad [15]$$

The products of Fenton reaction are hydroxyl free radicals which are extremely reactive and able to disrupt biological macromolecules like DNA, proteins, membrane lipids, etc. (Phillips et al. 2009; Barzilai and Yamamoto 2004; Simko 2007).

It is well known that the presence of oxygen enhances the action of free radicals within the cells, by reacting with them and forming more free radicals (Coggle 1983; Hall and Giaccia 2006). These oxygen-containing free radicals are called reactive oxygen species (ROS). In aerobic cells, ROS are normally produced by mitochondrial activity, (French et al. 2001; Barzilai and Yamamoto 2004). If ROS are not properly controlled by the cell, they can damage cellular macromolecules and especially DNA. The fact that oxygen which is an essential component of life can be at the same time so dangerous, is reported as the "oxygen paradox", (Barzilai and Yamamoto 2004). EMF exposure seems to be associated with free radical and especially ROS overproduction (Phillips et al. 2009; De Iouliis et al. 2009; Simko et al. 2007). In such a case, DNA damage may be expected.

Cells that are metabolically active (like the reproductive cells) or cells with a high concentration of free iron (like the brain cells) are expected to be more vulnerable to EMFs according to the above analysis. This is also supported by the experimental findings on reproduction decreases (Panagopoulos et al. 2004; 2007a; 2007b; 2010; De Iouliis et al. 2009; Agarwal et al. 2009) and the epidemiological findings on brain cancer induction (Khurana et al. 2009; Hardell et al. 2009; 2007). While glial brain cells may become cancerous after DNA damage leading to brain cancers, nerve brain cells do not divide and thus are not likely to become cancerous, (Mausset-Bonnefont et al. 2004). DNA damage on nerve brain cells may then lead to cell death or malfunction which are both linked to neurodegenerative deceases such as Parkinson's and Alzheimer's.

It is also possible that already existing free radicals and ROS, produced physiologically in cells, extend their life span in the presence of external EMFs.

Finally, another possibility for the biochemical mechanism is the combination of the above two described scenarios.

The above described ways of indirect biochemical action of EMFs on cells leading to DNA damage, seem to form a realistic basis for the biochemical explanation of the recently reported experimental results regarding DNA damage and cell death, induced by EMFs. Therefore, it seems that there is a plausible complete explanation (biophysical and biochemical) for the effects of mobile telephony radiations, reported in recent studies.

DOSIMETRY OF EMFs EXPOSURES. IS SAR A CREDIBLE QUANTITY?

While some studies refer to the radiation Intensity on the surface of the exposed sample in order to describe the exposure conditions, some others refer to the Specific Absorbtion Rate (SAR- the amount of energy absorbed by the unit mass of tissue). While radiation or field intensity can be readily and objectively measured, SAR is approximately estimated, usually by complicated numerical methods simulating living tissue by inanimate objects of similar shape and mass. In this section of the present chapter, an attempt is made to discuss the necessity of using or not SAR as a dosimetric quantity.

In all the above described experiments we referred to the radiation in terms of its intensity (at the distance from the antenna where the insects were exposed), which can be readily measured objectively, rather than in terms of *SAR*, which is not measured directly and can never be accurately estimated.

Usually *SAR* values are reported in papers without any information about the way of their calculation. Let us examine this quantity:

SAR is defined as the absorbed power P, per unit mass of tissue (Moulder et al. 1999; Panagopoulos and Margaritis 2003), (in W/Kg):

$$SAR = \frac{P}{m} \qquad [16]$$

where, $m = \rho V$, is the mass of the tissue of density ρ (in Kg/m^3) and volume V.

The energy density (in J/m^3) of an electromagnetic wave is given by (Panagopoulos and Margaritis 2003):

$$W = \frac{P \delta t}{V} = \varepsilon \varepsilon_0 E^2 \qquad [17]$$

where, W the energy per unit volume of tissue, transferred by the wave during a time interval δt, E the electric field component within the tissue induced by the wave (in V/m), $\varepsilon_o = 8.854 \times 10^{-12}$ C²/N·m² the dielectric constant of vacuum and ε, the relative dielectric constant of the tissue, (varying significantly for different tissues and different parts of a cell, for example $\varepsilon \cong 80$ for the aqueous solutions in the cytoplasm or the extracellular spaces and $\varepsilon \cong 4$ within membrane channel proteins).

By use of equation [17] and the Ohm's law

$$j = \sigma E \qquad [18]$$

where j, is the induced electric current density (in A/m²) within the tissue and σ the specific conductivity of the tissue, (in S/m), equation [16] after operations, becomes:

$$SAR = \frac{\sigma \cdot E^2}{\rho} \qquad [19]$$

For a homogeneous medium with specific heat c, [in J/(Kg·K)] and by use of a form of the heat transmission equation:

$$\frac{dQ}{dt} = m \cdot c \cdot \frac{\delta T}{\delta t} \qquad [20]$$

equation [16], becomes:

$$SAR = c \cdot \frac{\delta T}{\delta t} \qquad [21]$$

where: $\frac{dQ}{dt}$ is the wave power, transformed into heat, within the tissue of mass m, producing a temperature increase δT during the time interval δt.

SAR, is estimated by one of the following ways, (Moulder et al. 1999):

1) Insertion of micro-antennas within the tissue, which detect the internal electric field. If the conductivity and the density of the tissue are known, the SAR can be computed from eq. [19].
2) Insertion of miniature thermal probes within the tissue. If a change δT in the temperature of the tissue is recorded, caused by the radiation/field and the tissue is supposed homogeneous with known specific heat, then SAR can be computed by eq. [21].
3) Numerical modeling, like Finite Difference Time Domain, (FDTD) simulation, which simulates the spatial distribution of the radiation within a body.

Microwave energy when absorbed by matter, induces vibration on polar molecules and ions, superimposed on the thermal vibration of the same particles and therefore increasing their thermal energy. But the energy of the vibrations induced by external EMFs at environmental exposure levels, is thousands of times (about 10^4) smaller than the molecular thermal energy kT within a biological tissue, (Panagopoulos and Margaritis 2003; Panagopoulos et al. 2000; 2002). Thereby, EMFs at intensities encountered at human environment cannot cause thermal increases except if they were thousands of times more powerful, like for example the fields within a microwave oven which operates at about 1000 W in contrast to a mobile phone (~ 1 W) or a mobile telephony base station antenna (~ 100 W).

As it becomes evident from its definition, *SAR* expresses the rate at which electromagnetic energy from the external electromagnetic wave/field is converted into heat within a biological tissue (Stuchly and Stuchly 1996), therefore it assumes that EMFs bioeffects are exclusively related with thermal increases. But in our days it is well documented from many experimental studies (like the above presented experiments), that the biological effects of weak electromagnetic fields (at environmental levels) are non-thermal (Panagopoulos and Margaritis 2008; 2009), and plausible non-thermal mechanisms for the action of EMFs on cells are proposed as well (Panagopoulos and Margaritis 2003; Panagopoulos et al. 2000; 2002). This experimental evidence is in agreement with the above argument that environmental EMFs should be thousands of times more powerful in order to be able to induce thermal increases.

Additionally, since conductivity and density vary for different tissues and moreover, conductivity varies for different field frequencies, *SAR* varies also and therefore cannot be known accurately. Computer simulations, like FDTD method which is considered as the best way for computing *SAR*, dividing tissue volume into little homogeneous pieces (voxels) of constant conductivity and density can only be approximations. This is why earlier *SAR* estimations defining the current limits for whole body average *SAR* (ICNIRP 1998), are questioned by more recent and more accurate calculations, (Wang et al. 2006). On the contrary, the characteristics of the external field, (intensity, frequency etc.), can be measured, accurately. For these reasons the "exposure criteria" are given both in power density, (or electric and magnetic field intensities) and *SAR*, (ICNIRP 1998).

The necessity or not of the use of SAR as a dosimetric quantity, is a "burning" point. Whether or not someone agrees with the above analysis, we believe that we logically support our arguments and for this the above analysis may contribute to the debate on EMFs exposure dosimetry.

PROTECTION ISSUES. A POSSIBLE WAY FOR REDUCING RADIATION LEVELS WHILE MAINTAINING THE ABILITY OF COMMUNICATION

Mobile Telephony has undoubtedly become a part of modern daily life. It is useful because people can communicate at any moment from any place and it can even save lives in difficult moments. On the other hand, the exposure to its radiations may lead to serious health implications according to the experimental findings. The distinct similarity between some of the findings almost eliminates the possibility that these findings can be wrong, or due to randomness. Therefore people must be seriously educated in the schools about the dangers of using mobile phones - especially the children - and make very cautious use of these devices. Since the effects are shown to be cumulative (Panagopoulos and Margaritis 2010a), all users should drastically reduce the length and frequency of their phone calls to the minimum. Children, being much more vulnerable to radiation, should not use mobile phones, except for emergency situations. Mobile telephones must not be carried on the bodies unless they are turned off. This is because at the "standby" mode, every few minutes they emit a periodic signal lasting a few seconds, to maintain connection with the nearest base station antenna. These periodic signals are as powerful as the usual "talk signals" during a conversation. The users must make use of the mobile phone's loudspeaker and keep the handset at least 40 cm away from their heads and other most sensitive organs of their bodies like the heart or the reproductive organs, during conversations. All other ways of protection (like wire connected ear-phones – "hands free"), are less effective, or maybe even more dangerous (like the "blue tooth" devices) than the mobile phone itself, due to the existence of the intensity window described before. The mobile phone must not be held close to the head except for emergency situations.

The electronic circuits of the mobile telephony antennas of both mobile phones and base stations should be redesigned by the manufacturers in order to make the receiver circuits operate at max power while keeping the emitter's power to the minimum.

With regard to the involuntary exposure from base station antennas, a different network design should be attempted by the mobile phone industry. Instead of installing antennas everywhere within residential and working areas, they should perhaps try the following: Install powerful antennas on mountains and hills around the towns and a minimum number of low-power antennas at certain places within the towns at the largest possible distances from inhabitants, (in the middle of wide streets, on the roofs of the highest buildings with appropriate electromagnetic shielding on the roofs, within parks on appropriate towers etc). These low-power base station antennas within the towns should be just a little more powerful than a mobile phone, (5 W maximum output power instead of 100 W of the usual base station antennas). If there is sea or lake or even river adjacent to the town, then base station antennas could be installed on towers upon floating platforms as well.

Installing powerful base station antennas on the satellites could also be useful and complementary to the above ways.

Perhaps the engineers of the mobile telephony industry will argue that these proposals would not work, but what we propose here is reasonable and worth trying. Perhaps the signal will not be available everywhere by the ways we propose, but we think that safety is more important than signal availability everywhere.

The mission of technology is putatively to improve the living conditions of the human race. Technological evolution is accomplished by use of the natural powers, such as electromagnetism. But the use of these powers and the improvement of the living conditions must always be carried out with respect to the natural environment, and without undermining human health. Knowing the dangers of each new technological achievement and finding ways to use technology safely, might be even more important than technology itself.

REFERENCES

Adair R. K., (1991): *"Biological Effects on the Cellular Level of Electric Field Pulses"*, Health Physics, Vol. 61(3).

Agarwal A, Deepinder F, Sharma RK, Ranga G, Li J. (2008): *Effect of cell phone usage on semen analysis in men attending infertility clinic: an observational study.* Fertil Steril. 89(1):124-8

Agarwal A, Desai NR, Makker K, Varghese A, Mouradi R, Sabanegh E, Sharma R., (2009): *Effects of radiofrequency electromagnetic waves (RF-EMW) from cellular phones on human ejaculated semen: an in vitro pilot study.* Fertil Steril. 92(4):1318-25.

Aitken RJ, Bennetts LE, Sawyer D, Wiklendt AM, King BV. (2005): *Impact of radio frequency electromagnetic radiation on DNA integrity in the male germline.* Int J Androl., 28(3): 171-9.

Alberts B., Bray D., Lewis J., Raff M, Roberts K., Watson J.D., (1994): *"Molecular Biology of the Cell"*, Garland Publishing, Inc., N.Y., USA

Armstrong JS. (2006): *Mitochondrial membrane permeabilization: the sine qua non for cell death.* Bioessays. 28(3):253-60. Review.

Bacandritsos N, Granato A, Budge G, Papanastasiou I, Roinioti E, Caldon M, Falcaro C, Gallina A, Mutinelli F. (2010) *Sudden deaths and colony population decline in Greek honey bee colonies.* J Invertebr Pathol. Sep 23. [Epub ahead of print]

Balmori A, (2005): *Possible effects of electromagnetic fields from phone masts on a population of White Stork (Ciconia ciconia)*, Electromagnetic Biology and Medicine, 24(2): 109-119.

Barzilai A, Yamamoto K., (2004): *DNA damage responses to oxidative stress.* DNA Repair, 3(8-9): 1109-15. Review.

Batellier F, Couty I, Picard D, Brillard JP. (2008): *Effects of exposing chicken eggs to a cell phone in "call" position over the entire incubation period.* Theriogenology. 69(6): 737-45.

Baum JS, St George JP, McCall K. (2005): *Programmed cell death in the germline.* Semin Cell Dev Biol. 16(2):245-59.

Bawin, S.M, Kaczmarek, L.K. and Adey, W.R., (1975): *"Effects of modulated VMF fields, on the central nervous system"*, Annals of the N.Y. Academy of Sciences, 247, 74-81.

Bawin,S.M., Adey, W.R., Sabbot,I.M., (1978), *"Ionic factors in release of 45Ca 2+ from chick cerebral tissue by electromagnetic fields"*, Proceedings of the National Academy of Sciences of the U.S.A., 75, 6314-6318.

Belyaev IY, Hillert L, Protopopova M, Tamm C, Malmgren LO, Persson BR, Selivanova G, Harms-Ringdahl M., (2005): *915 MHz microwaves and 50 Hz magnetic field affect chromatin conformation and 53BP1 foci in human lymphocytes from hypersensitive and healthy persons.* Bioelectromagnetics. 26(3):173-84.

Bezanilla F., White M.M.and Taylor R.E., (1982): *"Gating currents associated with potassium channel activation"*, Nature 296, pp.657-659.

BioInitiative Report: *A Rationale for a Biologically-based Public Exposure Standard for Electromagnetic Fields (ELF and RF), (2007),* http://www.bioinitiative.org/index.htm.

Blackman,C.F., Benane,S.G., Elder,J.A., House, D.E., Lampe, J.A. and Faulk,J.M., (1980). *"Induction of calcium - ion efflux from brain tissue by radiofrequency radiation: Effect of sample number and modulation frequency on the power - density window".* Bioelectromagnetics, (N.Y.), 1, 35 - 43.

Blackman C.F., Kinney L.S., House D.E., Joines W.T., (1989), *"Multiple power-density windows and their possible origin",* Bioelectromagnetics, 10(2), 115-128.

Cavaliere V, Taddei C, Gargiulo G. (1998). *Apoptosis of nurse cells at the late stages of oogenesis of Drosophila melanogaster.* Development Genes and Evolution, 208(2): 106-12.

Chavdoula ED, Panagopoulos DJ and Margaritis LH, (2010), *Comparison of biological effects between continuous and intermittent exposure to GSM-900 MHz mobile phone radiation. Detection of apoptotic cell death features,* Mutation Research, 700, 51-61.

Chiabrera A., Bianco B., Moggia E. and Tommasi T., (1994), *Interaction mechanism between electromagnetic fields and ion absorption: endogenous forces and collision frequency,* Bioelectrochemistry and Bioenergetics, 35, 33-37.

Clark M.P., *Networks and Telecommunications,* 2nd ed., Wiley, 2001.

Coggle J.E., *Biological Effects of Radiation,* Taylor and Francis, 1983.

Creasey W.A. and Goldberg R.B., (2001), *A new twist on an old mechanism for EMF bioeffects?,* EMF Health Report, 9 (2), 1-11.

Curwen P and Whalley J, (2008): *Mobile Communications in the 21st Century,* In Harper A.C. and Buress R.V. (Eds), "Mobile Telephones: Networks, Applications and Performance", Nova Science Publishers, 29-75.

De Iuliis GN, Newey RJ, King BV, Aitken RJ., (2009): *Mobile phone radiation induces reactive oxygen species production and DNA damage in human spermatozoa in vitro.* PLoS One. 4(7): e6446.

Diem E, Schwarz C, Adlkofer F, Jahn O, Rudiger H., (2005): *Non-thermal DNA breakage by mobile-phone radiation (1800 MHz) in human fibroblasts and in transformed GFSH-R17 rat granulosa cells in vitro.* Mutat Res. 583(2): 178-83.

Drummond-Barbosa D. and Spradling A.C., (2001), *Stem cells and their progeny respond to nutritional changes during Drosophila oogenesis,* Dev. Biol. 231, pp. 265–278.

Eberhardt JL, Persson BRR, Brun AE, Salford LG, Malmgren LOG, (2008): *Blood-brain barrier permeability and nerve cell damage in rat brain 14 and 28 days after exposure to microwaves from GSM mobile phones.* Electromagnetic Biology and Medicine, 27:215-229

Everaert J, Bauwens D. (2007): *A possible effect of electromagnetic radiation from mobile phone base stations on the number of breeding house sparrows (Passer domesticus).* Electromagn Biol Med.26(1):63-72.

Foley K, Cooley L. (1998). *Apoptosis in late stage Drosophila nurse cells does not require genes within the H99 deficiency.* Development. 125(6): 1075-82.

Franzellitti S, Valbonesi P, Ciancaglini N, Biondi C, Contin A, Bersani F, Fabbri E., (2010): *Transient DNA damage induced by high-frequency electromagnetic fields (GSM 1.8 GHz) in the human trophoblast HTR-8/SVneo cell line evaluated with the alkaline comet assay.* Mutat Res. 683(1-2):35-42.

French PW, Penny R, Laurence JA, McKenzie DR. (2001): *Mobile phones, heat shock proteins and cancer.* Differentiation. 67(4-5):93-7.

Gavrieli Y, Sherman Y, Ben-Sasson SA. (1992): *Identification of programmed cell death in situ via specific labeling of nuclear DNA fragmentation .*J Cell Biol. 119(3): 493-501.

Goldsworthy A., (2007): *The Biological Effects of Weak Electromagnetic Fields,* Mast-Victims.org Forums / Health, **Apr 18**.

Goodman E.M., Greenebaum B. and Marron M.T., (1995*), "Effects of Electro- magnetic Fields on Mollecules and Cells",* International Rev. Cytol. 158, 279-338.

Grigor'ev IuG. (2003): *Biological effects of mobile phone electromagnetic field on chick embryo (risk assessment using the mortality rate),* Radiats Biol Radioecol 43(5): 541-3.

Gul A, Celebi H, Uğraş S., (2009): *The effects of microwave emitted by cellular phones on ovarian follicles in rats.* Arch Gynecol Obstet. 280(5):729-33.

Guler G, Tomruk A, Ozgur E, Seyhan N., (2010): *The effect of radiofrequency radiation on DNA and lipid damage in non-pregnant and pregnant rabbits and their newborns.* Gen Physiol Biophys. 29(1):59-66

Hall E. J. and Giaccia A.J., (2006): *Radiobiology for the Radiologist,* Lippincott Williams and Wilkins, Philadelphia.

Hamnerius I, Uddmar Th, (2000*): Microwave exposure from mobile phones and base stations in Sweden,* Proceedings, International Conference on Cell Tower Siting, Salzburg, p.52-63, www.land-salzburg.gv.at/celltower

Hardell L, Carlberg M, Söderqvist F, Mild KH, Morgan LL. (2007): *Long-term use of cellular phones and brain tumours: increased risk associated with use for > or =10 years.* Occup Environ Med. 64(9):626-32. Review.

Hardell L, Carlberg M, Hansson Mild K., (2009): *Epidemiological evidence for an association between use of wireless phones and tumor diseases.* Pathophysiology. 16(2-3): 113-22.

Hillebrand F. (Ed), *GSM and UTMS,* Wiley, 2002.

Honig B.H., Hubbell W.L., Flewelling R.F., (1986): *"Electrostatic Interactions in Membranes and Proteins",* Ann.Rev.Biophys.Biophys.Chem.,15.

Horne-Badovinac S and Bilder D, (2005): *Mass Transit: Epithelial morphogenesis in the Drosophila egg chamber,* Developmental Dynamics, 232: 559-574.

Hutter H-P, Moshammer H, Wallner P Kundi M, (2006): *Subjective symptoms, sleeping problems, and cognitive performance in subjects living near mobile phone base stations.* Occupational and Environmental Medicine; 63:307-313.

Hyland G.J., (2000): *Physics and Biology of Mobile Telephony,* Lancet, 356, 1833-36.

ICNIRP (1998): *"Guidelines for limiting exposure to time-varying electric, magnetic and electromagnetic fields (up to 300GHz)",* Health Phys. 74, 494-522.

Johansson O., (2009): *Disturbance of the immune system by electromagnetic fields-A potentially underlying cause for cellular damage and tissue repair reduction which could lead to disease and impairment.* Pathophysioloy. 16(2-3):157-77.

Khurana V.G., Teo C., Kundi M., Hardell L., Carlberg M., (2009): *Cell phones and brain tumors: a review including the long-term epidemiologic data,* Surgical Neurology, 72(3): 205-14.

King R.C. (1970), *Ovarian Development in Drosophila Melanogaster.* Academic Press.

Lai H, Singh NP. (1995): *Acute low-intensity microwave exposure increases DNA single-strand breaks in rat brain cells,* .Bioelectromagnetics, 16(3): 207-10.

Lai H, Singh NP. (1996): *Single- and double-strand DNA breaks in rat brain cells after acute exposure to radiofrequency electromagnetic radiation.* Int J Radiat Biol. Apr;69(4):513-21.

Lai H, Singh NP, (1997): *Melatonin and a spin-trap compound block radiofrequency electromagnetic radiation-induced DNA strand breaks in rat brain cells.* Bioelectromagnetics,18(6):446-54.

Lee KS, Choi JS, Hong SY, Son TH, Yu K. (2008): *Mobile phone electromagnetic radiation activates MAPK signaling and regulates viability in Drosophila.* Bioelectromagnetics. 29(5): 371-9.

Liman E.R., Hess P., Weaver F., Koren G., *1991: "Voltage-sensing residues in the S4 region of a mammalian K+ channel"*, Nature 353, pp.752-756.

Lixia S, Yao K, Kaijun W, Deqiang L, Huajun H, Xiangwei G, Baohong W, Wei Z, Jianling L, Wei W., (2006): *Effects of 1.8 GHz radiofrequency field on DNA damage and expression of heat shock protein 70 in human lens epithelial cells.* Mutat Res. 602 (1-2): 135-42.

Luukkonen J, Hakulinen P, Mäki-Paakkanen J, Juutilainen J, Naarala J., (2009): *Enhancement of chemically induced reactive oxygen species production and DNA damage in human SH-SY5Y neuroblastoma cells by 872 MHz radiofrequency radiation.* Mutat Res. 662(1-2):54-8.

Maber, J., (1999): *"Data Analysis for Biomolecular Sciences"*, Longman, England.

Magras, I.N.; Xenos, T.D. *RF radiation-induced changes in the prenatal development of mice.* Bioelectromagnetics 1997, 18, 455–461.

Mailankot M, Kunnath AP, Jayalekshmi H, Koduru B, Valsalan R., (2009): *Radio frequency electromagnetic radiation (RF-EMR) from GSM (0.9/1.8GHz) mobile phones induces oxidative stress and reduces sperm motility in rats.* Clinics (Sao Paulo), 64(6):561-5.

Markova E, Hillert L, Malmgren L, Persson BR, Belyaev IY. (2005): *Microwaves from GSM mobile telephones affect 53BP1 and gamma-H2AX foci in human lymphocytes from hypersensitive and healthy persons.* Environ Health Perspect. 113(9): 1172-7.

Mausset-Bonnefont AL, Hirbec H, Bonnefont X, Privat A, Vignon J, de Sèze R., (2004): *Acute exposure to GSM 900-MHz electromagnetic fields induces glial reactivity and biochemical modifications in the rat brain.* Neurobiol Dis., 17(3): 445-54.

McCall K., (2004): *Eggs over easy: cell death in the Drosophila ovary*, Developmental Biology, 274(1), 2004, 3-14.

Miller C, (2000), *An overview of the potassium channel family*, Genome Biology, 1(4).

Moulder, J.E., Erdreich, L.S., Malyapa, R.S., Merritt, J., Pickard, W.F and Vijayalaxmi, (1999): *"Cell Phones and Cancer. What is the Evidence for a Connection?"*, Radiation Research, 151, 513-531.

Navarro A. Enrique, J. Segura, M. Portolés, Claudio Gómez- Perretta de Mateo, (2003): *The Microwave Syndrome: A Preliminary Study in Spain,* Electromagnetic Biology and Medicine, 22 (2-3), 161-169.

Nezis IP, Stravopodis DJ, Papassideri I, Robert-Nicoud M, Margaritis LH. (2000). *Stage-specific apoptotic patterns during Drosophila oogenesis.* Eur J Cell Biol 79:610-620.

Nezis IP, Stravopodis DJ, Papassideri I, Robert-Nicoud M, Margaritis LH. (2002). *Dynamics of apoptosis in the ovarian follicle cells during the late stages of Drosophila oogenesis.* Cell and Tissue Research 307: 401-409.

Nikolova T, Czyz J, Rolletschek A, Blyszczuk P, Fuchs J, Jovtchev G, Schuderer J, Kuster N, Wobus AM., (2005): *Electromagnetic fields affect transcript levels of apoptosis-related genes in embryonic stem cell-derived neural progenitor cells.* FASEB J. 19(12): 1686-8.

Noda M., Ikeda T., Kayano T., Suzuki H., Takeshima H., Kurasaki M., Takahashi H.and Numa S., (1986), *"Existence of distinct sodium channel messenger RNAs in rat brain"*, Nature 320, pp.188-192.

Palmer L.G. , (1986), in *"New Insights into Cell and Membrane Transport Processes"*, (G.Poste and S.T.Crooke, Eds.), p. 331, Plenum Press, New York.

Panagopoulos DJ, Messini N, Karabarbounis A, Filippetis AL, and Margaritis LH, (2000*):* *"A Mechanism for Action of Oscillating Electric Fields on Cells"*, Biochemical and Biophysical Research Communications, 272(3), 634-640.

Panagopoulos D.J., Karabarbounis, A. and Margaritis L.H., (2002), *"Mechanism for Action of Electromagnetic Fields on Cells"*, Biochem. Biophys. Res. Commun., 298(1), 95-102.

Panagopoulos D.J. and Margaritis L.H., (2003), *Theoretical Considerations for the Biological Effects of Electromagnetic Fields,* In: Stavroulakis P. (Ed.) "Biological Effects of Electromagnetic Fields", Springer, 5-33.

Panagopoulos D.J., Karabarbounis A., and Margaritis L.H., (2004), *Effect of GSM 900-MHz Mobile Phone Radiation on the Reproductive Capacity of Drosophila melanogaster,* Electromagnetic Biology and Medicine, 23(1), 29-43.

Panagopoulos DJ, Chavdoula ED, Nezis IP and Margaritis LH, (2007a): *Cell Death induced by GSM 900MHz and DCS 1800MHz Mobile Telephony Radiation,* Mutation Research, 626, 69-78.

Panagopoulos DJ, Chavdoula ED, Karabarbounis A, and Margaritis LH, (2007b): *Comparison of Bioactivity between GSM 900 MHz and DCS 1800 MHz Mobile Telephony Radiation*, Electromagnetic Biology and Medicine, 26(1), 33-44.

Panagopoulos DJ and Margaritis LH, (2008): *Mobile Telephony Radiation Effects on Living Organisms,* In Harper A.C. and Buress R.V. (Eds), "Mobile Telephones: Networks, Applications and Performance", Nova Science Publishers, 107-149.

Panagopoulos and Margaritis LH, (2009): *Biological and Health effects of Mobile Telephony Radiations,* International Journal of Medical and Biological Frontiers, Vol. 15, Issue ½, 33-76.

Panagopoulos DJ, Chavdoula ED and Margaritis LH, (2010): *Bioeffects of Mobile Telephony Radiation in relation to its Intensity or Distance from the Antenna*, International Journal of Radiation Biology, 86(5), 345-357.

Panagopoulos and Margaritis LH, (2010a): *The effect of Exposure Duration on the Biological Activity of Mobile Telephony Radiation,* Mutation Research, 699(1/2), 17-22.

Panagopoulos DJ and Margaritis LH, (2010b): *The Identification of an Intensity "Window" on the Bioeffects of Mobile Telephony Radiation,* International Journal of Radiation Biology, 86(5), 358-366.

Phillips JL, Singh NP, Lai H., (2009): *Electromagnetic fields and DNA damage.* Pathophysiology. 16(2-3): 79-88.

Salama OE, Abou El Naga RM., (2004): *Cellular phones: are they detrimental?* J Egypt Public Health Assoc. 79(3-4):197-223.

Salford LG, Brun AE, Eberhardt JL, Marmgren L, and Persson BR, (2003): *Nerve Cell Damage in Mammalian Brain after Exposure to Microwaves from GSM Mobile Phones,* Environmental Health Perspectives, 111(7), 881-883.

Santini MT, Ferrante A, Rainaldi G, Indovina P, Indovina PL. (2005): *Extremely low frequency (ELF) magnetic fields and apoptosis: a review.* Int J Radiat Biol. 81(1):1-11. Review.

Simkó M. (2007): *Cell type specific redox status is responsible for diverse electromagnetic field effects.* Curr Med Chem., 14(10):1141-52. Review

Sokolovic D, Djindjic B, Nikolic J, Bjelakovic G, Pavlovic D, Kocic G, Krstic D, Cvetkovic T, Pavlovic V. (2008): *Melatonin Reduces Oxidative Stress Induced by Chronic Exposure of Microwave Radiation from Mobile Phones in Rat Brain.* J Radiat Res (Tokyo), 49(6), 579-86.

Stindl R, Stindl W Jr. (2010) *Vanishing honey bees: Is the dying of adult worker bees a consequence of short telomeres and premature aging?,* Med Hypotheses. 75(4):387-90.

Stuchly MA and Stuchly SS, (1996*): Experimental Radio and Microwave Dosimetry*, In: Polk C. and Postow E., (Eds), Handbook of Biological Effects of Electromagnetic Fields, CRC Press, 295-336.

Stuhmer W., Conti F., Suzuki H., Wang X., Noda M., Yahagi N., Kubo H. and Numa S., 1989, *"Structural parts involved in activation and inactivation of the sodium channel",* Nature 339, pp.597-603.

Tisal J., (1998), *"GSM Cellular Radio Telephony"*, J.Wiley and Sons, West Sussex, England.

Tomruk A, Guler G, Dincel AS., (2010): *The influence of 1800 MHz GSM-like signals on hepatic oxidative DNA and lipid damage in nonpregnant, pregnant, and newly born rabbits.* Cell Biochem Biophys. 56(1):39-47.

van Engelsdorp D, Hayes J Jr, Underwood RM, Pettis J. (2008) *A survey of honey bee colony losses in the U.S.,* fall 2007 to spring 2008. PLoS One. 3(12):e4071.

Wang J., Fujiwara O., Kodera S., Watanabe S., (2006): *FDTD calculation of whole-body average SAR in adult and child models for frequencies from 30 MHz to 3 GHz.*, Phys Med Biol., 51(17):4119-27.

Wdowiak A, Wdowiak L, Wiktor H. (2007): *Evaluation of the effect of using mobile phones on male fertility.* Ann Agric Environ Med. 14(1):169-72.

Weiss N.A., (1995): "Introductory Statistics", Addison-Wesley Publ.Co.Inc.

Yadav AS, Sharma MK, (2008): *Increased frequency of micronucleated exfoliated cells among humans exposed in vivo to mobile telephone radiations.* Mutat Res. 650(2): 175-80.

Yan JG, Agresti M, Bruce T, Yan YH, Granlund A, Matloub HS. (2007): *Effects of cellular phone emissions on sperm motility in rats.* Fertil Steril., 88(4):957-64.

Yao K, Wu W, Wang K, Ni S, Ye P, Yu Y, Ye J, Sun L., (2008): *Electromagnetic noise inhibits radiofrequency radiation-induced DNA damage and reactive oxygen species increase in human lens epithelial cells.* Mol Vis. 14: 964-9.

Zareen N, Khan MY, Ali Minhas L., (2009) *Derangement of chick embryo retinal differentiation caused by radiofrequency electromagnetic fields.* Congenital Anomalies, 49(1):15-9.

Zeni O, Schiavoni A, Perrotta A, Forigo D, Deplano M, Scarfi MR. (2008): *Evaluation of genotoxic effects in human leukocytes after in vitro exposure to 1950 MHz UMTS radiofrequency field.* Bioelectromagnetics. 29(3):177-84.

Zhang DY, Xu ZP, Chiang H, Lu DQ, Zeng QL., (2006): *Effects of GSM 1800 MHz radiofrequency electromagnetic fields on DNA damage in Chinese hamster lung cells,* Zhonghua Yu Fang Yi Xue Za Zhi. 40(3):149-52.

In: Mobile Phones: Technology, Networks and User Issues ISBN: 978-1-61209-247-8
Editors: Micaela C. Barnes et al., pp. 55-94 ©2011 Nova Science Publishers, Inc.

Chapter 2

MOBILE PHONES, MULTIMEDIA AND COMMUNICABILITY: DESIGN, TECHNOLOGY EVOLUTION, NETWORKS AND USER ISSUES

Francisco V. Cipolla-Ficarra[*]
HCI Lab – FandF Multimedia Communic@tions Corp,
ALAIPO – Asociación Latina de Interacción Persona Ordenador,
AINCI – Asociación Internacional de la Comunicación Interactiva, Italy

ABSTRACT

In our research work we focus on the users who have a greater difficulty in understanding the functionality of mobile phones with the triad of communicability, usability and usefulness. It is a detailed study which starts with the models of the 90s and the first decade of the new millennium, and tries to underline the area where the formal and factual sciences interact. To this end an unprecedented evaluation methodology is established where there is a technological and communicative intersection. The participation of real users allows us to put into use a series of heuristic research techniques to face these problems from several points of view of interactive and communicational design. In this work we make a special analysis of the subject of the choice of the random samples inside descriptive statistics, since we are working with it to cut down costs in the methods and evaluation of the quality in the interactive systems, where the mobile multimedia phones are included, for example. Additionally, the obtained results allow not only to know the weak and strong points of the current multimedia mobile phones but we set up a little vademecum where the sociological and technological aspects of the users in the coming years are summed up.

---

[*] HCI Lab – F&F Multimedia Communic@tions Corp, ALAIPO – Asociación Latina de Interacción Persona Ordenador (www.alaipo.com), AINCI – Asociación Internacional de la Comunicación Interactiva (www.ainci.com), Via Pascoli, S. 15 – CP 7, 24121 Bg, Italy – Email: ficarra@ainci.com

INTRODUCTION

Communication is an essential element in the human being inside a community since one tends to go from an individual communication to a social communication. The social communication media have tried to foster this phenomenon of socialization of people through the press, the radio, the television, the cinema, etc. in the last decades of the twentieth century. In this socialization process the phone has also been present, first landline and currently mobile and multimedia [1] [2] [3] [4]. However, the new technologies show a constant cycle of elements which act as centripetal or centrifugal forces according to the technological novelties which are included in the personal communication devices [5] [6] [7] [8] [9] [10] [11]. Many of these components are supposedly aimed at solving communication problems and quicken socialization among their potential users. However, in many cases these problems do not only persist, but they grow worse. It is an exponential worsening deriving from the passing of time and the potential users of the new technologies [12] [13] [14]. Currently we are witnessing a geometrical increase of the difficulties detected with the triad of mobile phones, multimedia and communicability deriving from factors such as ergonomic design, technology, the networks and the users [15] [16] [17] [18]. In other words, problems regarding communicability, usability and usefulness. In both cases the triadic components keep a bi-directional relationship among themselves. The magnitude of these problems is such that in some cases it can lead to non-communication among people, even with all the technological aspects working at a 100%.

In the mobile multimedia computing related to phones it is made clear how the user does not know the full 100% potential of the hardware and the software of the device in his hands [19]. This situation was common in the 90s in the context of the 2D and/or 3D designers. They used and knew part of the options of the commercial software and hardware aimed at entertainment and leisure time, for instance in the creations of computer animations for cinema and videogames [20]. The speed of novelties in hardware, new versions of software and the work to be developed prevented them from finding out the details of the tools with which they interacted daily. The same has happened with the evolution of mobile phones in Southern Europe. At first it meant the liberty to communicate from any place in the home or the office [21] [22], whether it was with spoken word or text messages: SMS (Short Message Service). However, little by little mobile phones have become multimedia instruments with which many senior users can't easily communicate, remaining totally sidelined upon the arrival of the new phone models. Those who know some of their basic functions have again the same problems after a short time because they are forced to change the phone because the batteries over time loose capacity, and they are no longer made. Besides, in Europe there are no training centres to teach how to use multimedia phones for users with difficulties.

Consequently the apogee of the mobile phone which has led to many countries in Southern Europe to have the highest fees in landline communication with the alleged purpose of boosting the phone communication network, has left part of their population cut away from these breakthroughs, and sidelined technologically speaking. For instance, Telefónica from Spain had the most expensive fee between Lisbon and Moscow in the first half of the 90s. Obviously the cost factor and low purchasing power has also prevented some users of mobile phones from updating and getting acquainted more closely with new models to not stay in the technological rearguard. To this contextual factor we may add other realities from France or

Italy where many people usually have more than one mobile phone in order to split work, family and leisure time, for instance. Some of them use the same model of mobile phone, others, in contrast, prefer different models in order to always be acquainted with the latest technological novelties which fashion thrusts upon them.

Using more or less mobile phones simultaneously is not synonymous with greater social communication as it happened with the fruition of the traditional media: movie houses, family gatherings to listen to or watch a soap opera on radio or television in the last century. The multimedia mobile phone user tends to participate unwittingly in the autism phenomenon in the current era of global communications. Oddly enough, the advantages of freedom and socialization of the individual entailed by the communication media disappear, because of the complexity of the access to the contents and the degree of complexities or functions of it among given potential users [23] [24] [25].

The current work is structured in the following way: a constant diachronic analysis of the new technologies focused on computer science, telecommunications and the communicability of users with mobile phone systems and other interactive communication systems, in the daily life of their activities, description of the techniques, methods and strategies used deriving from the intersection of the formal and factual sciences, with special emphasis on the correct realization of the samples of study, bearing in mind the details of descriptive statistics: study of the main software and hardware characteristics of the mobile phones, in the past, present and with a particular look at the future; presentation of the obtained results and future lines of research.

TELEPHONE EVOLUTION, MASS MEDIA, COMPUTER SCIENCE AND GLOBAL VILLAGE

Since the dawn of the telegraph, the radio and other technological breakthroughs of telecommunications throughout the 20th century, there has always been an attempt to unite the territory of a nation and its inhabitants. That is, it was an implicit and/or explicit goal of a long series of discoveries and inventions in telecommunications until reaching the current global village, where the mobile phones and the Internet have an overruling role in the social evolution and the economic development of societies [26]. First, it was the audio in the radio, then the radio was joined to the image with television, until the appearance of interactive communication with the Internet in the mid 90s and then the multimedia mobile phones. In an early stage of social communication it was usual to go from one situation of predomination of the emitter over the receptors, since there was no possibility of interaction with the media, as it was the case with analogical radio and television.

Later on, the democratization of the use of the PC connected to the Internet broke with that verticality in the communication process towards a horizontality and the promotion of the free access of contents for everyone in the early 90s, first with the computers and now with mobile phones. In the new millennium, one can see situations of a return to a position of dominance by some members in the social groups through persuasion. It is the case of the dynamic persuader and interactive persuaded person [27] in Web 2.0, for instance. An individual who is capable of breaking up the democratic horizontality of the web for his/her own benefit or for the benefit of an elite to which he/she belongs. This is achieved through the constant manipulation and persuasion made among the members of his social networks. That is, he generates small spaces of vertical communication and with unethical and dictatorial behaviours. Mc Luhan's global village based on the democratization of information is being attacked by the behaviour of the dynamic persuader, who moves freely inside the social networks [27]. This phenomenon is much more serious than that foreseen by those who claim that the Internet was invented for purely military reasons.

Many of these behaviours are related by some to the origins of the research projects in the new technologies, that is, the military context, such as have been the first text messages sent via the net [28]. However, in the ARPA-Net project, which was born with the purpose of joining the university researchers it has nothing to do with the use of the net for purposes of national security which was later applied to it. In the context of the mobile phones there are those who see the radios used in the Second World War (some models had a phone pipe) as the origin of its diffusion. However, it is necessary to remember that the radio and the phone are two very different entities. The phone is a telecommunication device designed to transmit acoustic signals through a long distance. It is very similar to the teletrophone. From the technical and historical point of view, the teletrophone is the name that was given by the Italian inventor Antonio Meucci to his version of the phone in 1870, who had built the first electromagnetic phone in 1856 [29]. Alexander Graham Bell was the first in patenting it, in 1876, in contrast to Meucci, who didn't have the financial resources to do it in the U.S. Patents Office and who could only hand in a short description but couldn't formalize the patent because of his lack of financial resources [30]. This dispute between both inventors of the phone was solved in 2002 when the U.S. congress passed resolution 269, which recognized that the inventor of the phone had been Meucci and not Bell.

In our case, the mobile phones, also called cell phones, are basically made up by two main parts: a communication network (or mobile phone network), and the terminals (mobile phones) which allow access to said network. Here it is interesting to see how the word "cell" or "mobile" divides from the linguistic point of view several countries in the Mediterranean basin, and the countries in the American continent. Just to mention a few examples, they are called mobile in Spain, Portugal, etc. and cellular in Italy, Argentina, Chile and Uruguay. Now the mobile phone is a wireless electronic device which allows to have access to the net of cellular or mobile phones. It is known as cellular due to the masts that make up the net, each one of which is a cell, although there are satellite mobile phone nets. Their main feature is their portability, which allows one to communicate from almost any place. Although its main function is voice communication, like the conventional phone, its fast development has incorporated other functions, such as a photo camera, notebook, access to the Internet, video reproduction and even GPS (Global Positioning System) and MP3 player. Obviously, another of the factors to be considered at the moment of speaking of the evolution of the mobile phones is the weight of the first non-commercial devices which surpassed the kilos as compared to the current ones which have the weight of just a few grams, like those which have the shape of a wrist-watch.

Multimedia communication has been boosted through the use of mobile phones [31]. Obviously, this has been a real revolution in the interactive communication of the users. The first forerunner of the mobile phone was from the Motorola company, with its Dyna TAC 8000X model. The model was designed by Rudy Krolopp of the Motorola company in 1983. The model weighed little less than a kilo, and had a market value of almost 4,000 US dolars. Krolopp would later join the research and development Motorola team lead by Martin

Cooper. Both Cooper and Krolopp appeared as owners of the original patent. From the DynaTAC 8000X onwards, Motorola developed new models such as the Motorola Micro TAC, launched in 1989, and the Motorola StarTAC, launched in 1996 to the market.

The design of the first mobiles mainly emulated the dimensions or functions of the traditional or landline phones (figure 1). However, little by little, as has also happened in the technical evolution of computing, a diminution of the size of the telecommunication device has taken place. Obviously, this design factor has also joined the usability and communicability factor of the first mobile phones. In these the functions were limited to the utmost, there were no color screens, nor multimedia functions, for instance. In some way there was a bid for the bigger public to get acquainted with the possibility of being able to call by phone without using a public booth when one was outside the home or the work office. Until this point such devices existed in the collective imagination of a society through the American television series. Where the mobile phone was joined to a high priced car of in a spy's shoes, like Maxwell Smart in the 1965 'Get Smart' TV series [32]. That was already a reality in the hands of millions of users. Later on, the costs of the phones started to decrease, the same as happened with the professional computers, which went on to become personal computers until in less than a decade in some economically developed countries almost every home has it available as if it were a Digital Terrestrial Television (DTT or DTTV) or a radio.

Figure 1. First commercial mobile phones in Spain and Italia –Motorola, Alcatel and Sony Ericsson (1991, 1998 and 2005).

Both in mobile phones as in the personal computers there is a tendency to diminish the size of the hardware (CPU, peripherals, etc.), and of the programs for their functioning. This tendency leads to the currently taking place parallelism between the personal computers and the multimedia phones. Evidently, if we consider the breakthroughs in the new technologies as a pyramid we have in its upper part the R+D area and the sectors of the population with a greater purchasing power. Whereas in the lower parts of the pyramid are the users who push for free access to information and the last technological advances with the least possible cost. Currently it is easy to detect how in the different communities the university and/or industrial labs are orienting their works only at the summit of that pyramid., especially in the great cities of the European Mediterranean basin, such as can be Rome or Barcelona. It is easy to find in those places because of fashion motives the use of mobile phones in a watch. Some models are GSM (Groupe Special Mobile) Triband, with double SIM (Subscriber Identity Module), photo camera and video (figure 2).

Figure 2. Mobile phones –Motorola, only SMS (1991) and Samsung SGH-L170–Multimedia phone (2008).

Analyzing some of its main characteristics, we can see how the microphones and the microcomputers converge on the wrist of the arm of a user under the function of a watch: a color screen 1.3", TFT (Thin Film Transistor) 360K, touch screen 128*160, 1,3 MPixel digital camera, MP3/MP4/3GP play, storing of JPG/GIF photos, Dual Stereo Bluetooth, spot for T-Flash card with memory until 2 GB, GSM, /GPRS/ WAP, videogames, alarm function, email POP3/SMTP, battery duration for conversations between 3 or 4 hours, 64 ring tones, etc. All of this with a 61x45x16 mm and a weight of 75 grams. To this device we can connect a Bluetooth battery load or a network adapter, for instance. These are the peripherals which are pacing the rhythm between the purchase of computers or multimedia mobile phones, since both started decades ago the miniaturization process of the CPU and the peripherals.

From the point of view of communicability and usability we find two main elements which may lead the adult users to purchase mobile phones instead of small computers (Pocket PC, Tablet PC, etc.), the VLK (Virtual Laser Keyboard) and the new generation of flexible screens (figure 3).

Figure 3. Mobile phones and micro information and communication technologies: iPhone (1), VLK (2), BlackBerry (3), Watch and Chronophone Mobile (4), and OLED sheet (5).

- The VLK allows to use a smartphone or PDA as a normal computer, with an enlarged keyboard, even in the dark. This device projects a QWERTY keyboard on a flat surface and allows to digit the keys as in a normal computer. Its main characteristics, which may vary among the different models, can be summed up in the following way: laser source, over 60 keys keyboard in full format QWERTY. Size of the virtual keyboard 295x 95 mm. Possibility of recognizing up to 400 characters per minute via Bluetooth. It usually works with a lithium-icon battery. The system is compatible with PalmOS 5, Pocket PC 2003, Windows Smartphone, Sybian OS, Windows 2000/XP.
- The new generation of flexible screens are made from an organic semiconductor material with transparent electrodes. Which allows them to be bent and to be very thin (almost of the thick of a hair). The same as with the early CDs, it is a technology patented by Sony, who has not announced yet to what products it will be applied. The Japanese brand Kyocera is also developing EOS, a flexible mobile phone which looks like a wallet. Nokia has also developed a prototype, the 888 model, a spectacular mobile phone with the shape of a wristband. The lightness of these

screens will make it possible to place them on almost any object, which makes the only limit to be creativity or imagination, especially in the multimedia sector: newspapers with flexible OLED (Organic Light-Emitting Diode) sheets, soda packages wrapped in this technology that shows ads, or television sets which will be no thicker than a poster hanging on the wall.

Once again the peripherals set trends among the different kinds of users and the different reasons for which one interacts with the interactive systems. Evidently, the goal is that the multimedia phones go down to the basis of the pyramid as soon as possible to boost the expansion era we are going through. Now in regard to the communicability it is necessary to differentiate the aspect of the design of use of the interactive systems which belong to the classical environment of computers usability and in this case it would be microcomputing applied to the telecommunications systems. In our work methodology, which has been developed for years in several sectors such as education, tourism, e-commerce, etc., one of the design categories of the interactive systems is connectability. This notion allows to differentiate what is traditionally associated to the hardware and software such as is the practical acceptability stated by Nielsen from system acceptability [33]. That is, he contends that the acceptability of a system is made of the social acceptability (without developing its content), and the practical acceptability, in which we have cost, compatibility, reliability, usefulness (in which usability is inserted and its five classical principles efficient to use, ease to learn, few errors, easy to remember and subjectively pleasing [33]. In contrast, compatibility is another component to be considered at the moment of designing an interactive system, especially in the current competition between computers and mobile phones. By correctly dividing the design in the interactive systems in several categories it is feasible to reduce their production costs, since it is possible to quickly detect where are the failings in the realization stage of the prototypes of the mobile phones, for instance.

COST FACTOR, TELECOMMUNICATIONS DIFFUSION AND DESIGN ASSESSMENT

The cost factor in the purchase of the new technologies by the basis of the pyramid: the population is essential for the expansion of communicability, where the dynamic and static means of interactive communication play an essential role. In the case of phones, we can see how the devices have been adapted first for the text messages and then little by little for the multimedia. At the start the mobile phone was used for conversations and avoiding the possibility of sending SMS messages [34]. Today, some mistakenly claim that the emails and the SMS tend to disappear as a result of the apogee of the web. However, the text in the communications is the basis of the hypertextual communications and will remain because when confronted with doubt in the veracity of the dynamic and static images the human being tends to believe in the written word.

There were two reasons: the costs of the phone reloads and the lack of experience of writing on such a small device for millions of young and adult people. In regard to the cost of the service it was very high in the 90s in the countries of the Mediterranean basin. This was due to the monopoly of the state enterprises in the telecommunication field which forced the users at the moment of the mobile phones reloads to pay additional sums for the simple fact of making a reload ("ricarica"), as it can be seen in the following image of the Italian mobile phone cards (figure 4). That arbitrary plus of the phone companies was in use until 2008, where the private companies appeared in Italy. In the Spanish case, the high costs were due to the VAT percentage applied to the reloads ("recarga"). In others words, since the times of the peseta in Spain additional services were never paid for reloading the mobile phone (Telefonica –MoviStar). In contrast, in Italy, at the time of the lira, the directors of the public and private companies made people pay for this service "ricarica", even with the arrival of the euro (Telecom –TIM).

Figure 4. Evolution of the prices in the phone cards for mobiles between Italy (TIM) and Spain (MoviStar), including the swap of national currency (peseta and lira) towards the euro.

Even these enterprises in the prevailing position of the mobile phones achieved financial records such as those reached by the Telefonica phone company. They got an extra 40% profit in the 2004-2005 fiscal year, (over 5 billion euros). However, with the passing of time and the democratization in the spread of these devices among the populations of the economically developed EU countries, these abusive costs of the mobile phones were accepted by almost everyone as a way of paying for the additional advantages that were inherent to the possibility of freely communicating from any place in the planet. That is, a kind of apocalyptic-biblical principle applied to technological innovation: before enjoying the technological breakthroughs it is necessary to suffer.

Now the cost problems and the way of managing the origin of the diffusion of mobile phones expanded through several countries in the South of the American continent. However, many of these communities are used as labs of the new technologies to be later on applied to the developed countries. Even in trivial issues such as the marketing of mobile phones. In the first years of the new millennium more mobile phones were sold than could actually work because of the infrastructure that existed in those places. A structure of telecommunications that were developed and managed by the very same firms of mobile phones, with their headquarters in Spain, France and Italy. The real users of mobile phones were submitted to real tolerance tests concerning the malfunction of mobile phone services and which later on were applied to the diffusion and marketing of the Internet. All these studies of public reaction or of users have served to automate mobile phone services, landline, Internet, etc., in Europe.

For instance, with the programming of the virtual assistants to solve the technical problems of Internet connection using mobile phones. In some regions known as traditionally as the engines of Europe such as Baden-Württemberg, Catalonia, Rhône-Alps and Lombardy it is possible to see how mobile phones are used to cheat the final user and increase the costs in the internet services. For instance, in Lombardy it may happen that the subscribers to the phone lines surpass the available lines for the high speed internet services, consequently there are subscribers who do not have the service continuously available, as it happens in many developing countries. That is, cyclically some subscribers are continuously disconnected from the internet. To solve this problem, the subscriber activates a second or third line of high speed internet of high speed to have the internet services continuously available. Automatically he/she will keep on receiving fake SMS on his/her mobile phone which say that his/her service has been activated (figure 5), when in fact, 10, 20, 30 days may pass without Internet service, and having to pay for a non-existing service. Additionally, the virtual assistant will communicate that the failing is due to problems in the central phone service. Consequently, there are cities inside the so-called engines of European economy where the user is deceived with SMS and has to multiply by two or three the costs of the phone and internet services. This reality slows down the diffusion of mobile phones in a wide sector of the European population, for instance Aside from these negative realities for the progress of the new technologies in Europe, most users started to invest in the new technologies inside telecommunications in the 90s in a cautious way because of the novelty, plumping for the basic functions, that is, those that were cloned in the new devices and which offered that additional freedom of movement for city telecommunications, inter-city and international (late 90s). Then the public started to use the phones for sending SMS during almost the whole first decade of the new millennium, until the multimedia mobile phones allowed them to participate in the social networks from his/her mobile [35]. This is an interesting area to be analyzed in the future, whether it is from the sociological and/or social psychology point of view, seeing the negative impact among the Human-Computer Interaction, the communicability and the users with the traditional and multimedia mobile phones. In the diffusion of mobile phones not everything is focused on communicability and usability of the new devices, but in the human and economical factors which are included in the new technologies. Consequently, it is important to approach these problems from the convergence of the formal and factual sciences, for instance.

Figure 5. Italian fake SMS about Internet service (ADSL) on mobile phone.

HEURISTIC EVALUATION: TOWARDS A RELIABILITY OF THE RESULTS

At the beginning of the mobile phone era, the quality demanded by the users focused on the usability of the device and the biggest possible coverage of the mobile in the phone network. With the passing of time and the incorporation of the dynamic and static means in the mobile phones, usability was replaced by communicability for the user himself/herself. Later on, communicability requisites have been increasing as the multimedia mobile phones could be wired to the net and use the last novelties deriving from the Web 2.0. Along the time a series of heuristic evaluation techniques have been developed. However, there wasn't in the first years of the mobile phones an evaluation methodology of interactive design. With the incorporation of the dynamic and static means, the mobile phone became a multimedia device entailing an endless variety of design of the interactive systems of the computers. In this regard, a set of techniques and methods of heuristic assessment for the multimedia system, such as: MEHEM (MEthodology for Heuristic Evaluation in Multimedia) and MECEM (MEtrics for the Communications Evaluation in Multimedia) [36] [37] [38]. This set was based on a set of quality attributes deriving from HCI, software engineering, interfaces, primitives of design of hypertextual systems, multimedia and hypermedia, and the semiotics or semiology. These quality attributes have been increasing along the years in regard to the different applications and/or heuristic assessments which were carried out for education, tourism, e-commerce, etc. [39]. Obviously, these are quality attributes that can be broken down into metrics to assess the quality of the interactive systems. An assessment of the whole system or part of it, with the results always 100% reliable, carried out by evaluation experts, and verified in some cases in usability laboratories. The goal was to establish quality attributes validated by other heuristic evaluation techniques, such as videotaped sessions, guided interaction, beta-testing, user feedback, user surveys, focus groups, etc.

After the establishment of the set of quality attributes for the design of the multimedia/hypermedia systems there is a set of them which are related with a greater interdependence intensity or "subordination" among themselves. This interdependence makes it possible to elaborate a sketch of those attributes which have a higher degree or level of interdependence. For instance, the factual function is inside or derives from accessibility, or the motivation depends on the wealth or what is now called "augmented reality". A grouping of such attributes is to be found in the figure 6.

The use of heuristics for the generation is due to the fact that they respond to the following factors enunciated by Nielsen [33]: Speed, ease of implementation, and low cost. Although as a denomination some of them exist inside software literature, the here presented attributes are aimed at the multimedia systems in mobile phones. The heuristic attributes which have been presented are not orthogonal among themselves, but they are based on a communicability model and in a procedure which has been in a continuous definition, correction and verification. The use of a terminology stemming from the design models in the on-line and off-line interaction systems, semiotics or semiological, the primitive actions of interactive communication and in a special way the evolution of the hypertext, multimedia and hypermedia, allows us to hone even more the definitions although sometimes we resort to a broad terminology. For instance, the notion of the richness in which it is necessary to set down the limits of its scope.

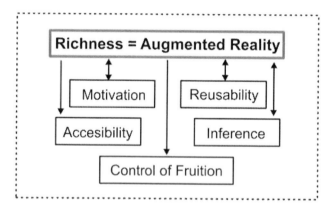

Figure 6. Richness and the quality attributes relations.

The establishment of the accurate limit of the significations is not easy, especially when one works with techniques which stem from the social sciences, such as is the case of heuristic assessment. In regard to the richness attribute, which was presented for the first time in 1999 [40], together with other quality attributes, it has been derived to the current notion of augmented reality [41] [42] especially in the interfaces at the moment of navigation, when the user can decide to incorporate or eliminate resources, or change the modalities of visualization of the information, for instance, in Google maps with its traditional options to locate a street or a city on the different maps, that is, Earth, satellite and street guide. However, there is no reference to the criteria of quality of richness or the accessibility (figure 5), another criteria of quality presented before finishing the past century [40] [43]. In the context of interactive design pioneers are often unfairly forgotten by resorting to the use of synonyms to define the concepts. The reason in these cases is the personal competence by certain clans inside the university sector in the Southern Mediterranean and for economical purposes, since the authors can present their works in publications of lesser international circulation, such as may have other publications of international associations, related to computing or electronics, for instance.

However, with such attributes one intends to set down the basis of a previously unseen methodology in the evaluation of the interface of a multimedia/hypermedia system. Their importance is even greater because there is still no "guidebook" to get 100% success on designing an application. The solutions that are suggested still require high investment, such as the creation of labs and counting on plenty of human resources [33] [39] [44] [45]. Once again the cost factor demands thinking of solutions which take into account the high quality equation of the results plus the least possible time and with the least cost. Under these conditions, it is necessary to find the best possible solutions derived from the social sciences and descriptive statistics, such as can be the straight observation techniques and the use of samples with the lowest possible error in the making of the universe of study.

HEURISTIC EVALUATION: ACHILLES HEEL

The research in the field of the factual sciences and where many of the heuristic sciences of evaluation of usability engineering confirm a kind of subset of the former. Social research, as Mario Bunge [46] and Ezequiel Ander-Egg [47] say, is a reflexive, systematic and critical procedure. This entails the need of organizing beforehand the whole research process and planning its execution in several stages. The organization and programming of research consist in projecting work in keeping with a logical structure of decisions and with a strategy that gives an aim to the way of obtaining the adequate data to the planned objectives. Neither the planning nor the stages to be followed by the evaluators are not described in those works related to the education and the heuristic evaluation of the interaction of the young users of new technologies in Catalonia or in Lombardy. Perhaps the word "planning" has implications that are not liked by the mentors of these works. In contrast to this modus operandi, a correct planning is indispensable to cheapen the research costs. In the current work the notion of organization includes the scientific and technical aspects of the different stages and phases which must be implemented in order to carry out a research and which in our case can be summed up in the following way:

1. Determination of the main and secondary goals to be reached.
2. Exploratory phase. Consult and documental compilation.
3. Structuring of the research design: elaboration of a theoretical framework, shaping of a team of evaluators and users, task coordination, evaluation techniques and methods, organization of the consult and research material, determination and choice of the sample, management of the financial resources to cover the expenses of the lab (use of computers, payment to the users, etc.)
4. Preparation of the work group.
5. Post-evaluation work: classification of the obtained results, analysis and interpretation of the data and/or information, writing of the final reports.

In the organization and planning of the research work in the social context one can start with the classical Platonic rhetoric questions and then add others. That is, the questions that we ask ourselves at the beginning of the work and the goals we must set. In this regard the following parallelism can be established: what (issue or subject), what for (goal), why (problem), where (area of the factual and/or formal sciences), when (activities chronogram) how much (intensity degree of the study, for instance, the whole universe or a part of it, that is, a sample), how (methods and techniques), who (human team), with what (technological resources and implicitly economical). In few words, it is about making operational the scientific method applied to a given field of research.

Currently there is no guide within the usability and usefulness concept which can serve us as a kind of compass in the face of the problems to be solved and all the decisions that must be made in order to carry out such research. In view of such a situation we will also try to include communicability, usability and usefulness of the mobile phones in this kind of guide where are indicated the stages or logical steps of the research process. It is necessary to make clear that these stages or steps which are taken in the context of the social sciences and in our case, by working with real users, some of the here indicated stages may overlap with other analogous situations where communicability is not considered. Now it is important to point out that the organization of the research work in the heuristic research environment including its methods and techniques in the formal and factual sciences consists in making available all the necessary operations for its realization. Carrying out a research task without an adequate preparation as may be the case of a usability expert but who is inexpert in communicability may require a longer time than what is actually needed. In well-prepared research, there is neither hurry nor unnecessary waste of time in the preliminary tasks, which in some cases entail high costs in relation to the profits or obtained results. In these preliminary tasks we can insert those communicability elements that the team of experts, jointly with the users have to locate regarding the interaction of the mobile phones system and which need an explanation or previous example.

Formulation of the aspects to be researched. In this context it is important to determine what is the goal of the research, the purpose and the delimitation of the field to be researched. Everything has its beginning in a good formulation of the problem that delimits the research and is its target. Once the problem has been formulated it is necessary to subdivide into several subsets or sub problems, as many as possible. This operation includes the task of explaining the aspects, factors or relevant communicability components, related to one or several categories of the design that is the aim of the research, for instance. Once the main goal or problem of the aspects to be researched has been formulated in an accurate and operational way, the evaluator will start to solve each one of these sub problems until encompassing them in their totality. It is important to point out that each one of these subproblems inside the scientific methodology are denominated dimensions and variables of the fact to be researched. For instance, when the problem of evaluating the design of the interactive systems was posed, it was divided into several subcategories to facilitate its resolution in the creation of methodologies, techniques, instruments, etc., and bearing in mind several factors related to the potential users and their cultural aspects, among others. Once again the rhetoric questions may help the evaluator as it is our case. Starting from the fact that the ability to formulate problems is something that is related to the scientific talent, some rhetoric questions that may help develop this talent are:

- What is the problem? What are the data of the problem? What are the main aspects or elements of the problem? (identification of the problem in a clear and accurate way, thus decomposing it into its components, establishing categories and variables or dimensions).
- Is it an unprecedented problem or not? (to research the existing bibliography in the issue that will be approached)
- Which are the relationships among the different components of the problem? Which are the issues related to the problem? (translation of the approaches into variables that can be measured).
- Has lexical ambiguity been eliminated? (a clear and accurate conceptual framework must be generated so that then the different agents that will intervene in the later stages do not generate confusion and waste time. Wasted time is equal to economical resources).
- Has the finality been defined? (each one of the operations that will be made in the evaluation must lead to solving the main problem).

Consequently, a correct formulation of the field to be analyzed or examined must respond in a clear and accurate way to the what and why of the evaluation. Additionally, the evaluation field must be limited, for instance to a whole universe of study or a sample. In this latter case, the mechanisms must be established through which the elements are randomly chosen that will be researched to guarantee the greatest possible objectivity in the results which are later obtained. In our case, we have had to define the groups of adult users and the models of mobile phones which were used in the experiments made. That is, several types of users divided according to age, studies, previous computer knowledge, etc. and several models of phones classified by decades, functionalities of said phones, presence or absence of the dynamic and static means, etc. (our users were to be found in the cities of Barcelona – Spain, and Bergamo –Italy, and the remarks made refer to the models of mobile phones which go from 1998 until 2010). Once this phase has been left behind, we have to analyze the existing references in relation to our problem, bearing in mind taking into account the originality of the research and problems resolution process. In this exploratory stage, the review of the existing literature is of great importance. In the social sciences, some authors [47] [48] contend that the successive stage is called research design or tasks to be carried out, that is, a kind of outline in which is indicated the set of activities to be carried out to steer the course of a heuristic evaluation of an interface in a multimedia interactive system, for instance. Therefore, we are going full steam into a set of tasks which can be grouped in the following way: elaboration of a theoretical framework, confirmation of a work team (especially in the usability labs). In our case, the communicability expert does not require that work team, however, it can be made up to verify the obtained results through some heuristic evaluation techniques where users intervene, for instance, interviews, video sections where the interaction with the mobile phones is recorded, etc. Continuing with the eventual need of working with a researchers and/or users team, it is necessary to coordinate the tasks of the researchers or project advisers, the surveys and stats team (if there were), computing and telecommunication experts (technical issues in the interaction process with the mobile phone models), auxiliaries for the data compilation and writing of partial reports, etc. Then there are the selection stages of the methodological instruments, for instance evaluation tables, guides, discs, etc. the organization of the consultation material for the members of the evaluating process, choice of the field of study (total universe or sample).

SAMPLE FOR HEURISTIC EVALUATION

The meaning of the sample designates the part or representative element of a set. If we use the term in an exact sense, as it is used in statistics, the sample is part of a population which is the target of a research. For instance, the users in our case stemmed from professional training courses and they were taking computing courses for resinsertion in the labor market. Others, in contrast, attended language courses and were handpicked following each one of the steps which will be analyzed later on, in order to obtain a random sample in the strict sense of the notion. This sample is chosen by certain procedures, and its study leads to conclusions which are extensive to the whole of the members of the set with a significant economy of cost and with a greater speed of execution. Now, when one chooses to use samples instead of the whole of the universe of study, said samples have to be planned and carried out in a careful way since the value of the conclusions depends on the representativeness of the sample. From the conceptual point of view, in the sampling process are used a series of basic concepts which are next described to eradicate ambiguity [47]:

- The sampling technique is the set of operations which are made to choose a sample.
- The unity of the sample is made up by one or several elements of the integrants in which is subdivided the basis of the sample and which inside it are defined in a very accurate way.
- The basis of the sample is the set of individualized units that make up a universe of study.
- The universe of study is made up by the whole of a set of elements that are being researched and from which a fraction or sample will be studied which is expected to gather the same features and in the same proportion.

Some authors distinguish between universe and population [48] [49], that is, with the first term a set of elements is denominated, beings or objects, whereas with the second a set of obtained numbers by measuring and counting certain features of said objects. In the current work and with regard to communicability we prefer to use the first notion, although we regard both as equivalent or similar [39]. Although it is true that we are examining real users at the moment of the interaction with the mobile phones, that is, population belonging to one or more communities, the notion of universe of study is more correct in the context of communicability.

Now the sampling method is based on a set of laws that provide it with scientific foundations, that is, the law of large numbers, and calculation of probabilities, excepting the case of empirical or non random samples. The law of large numbers, formulated by Jacques Bernouilli, is expressed in the following way [49]: If in a test the probability of a happening or event is 'p', and if this is repeated a great number of times, the relationship between the times in which the event takes place and the total amount of times, that is, the 'f' frequency of the event, tends to get closer every time more to the probability 'p'. More exactly, if the number of tests is sufficiently big, it is totally unlikely that the difference between 'f' and 'p' surpasses any predetermined value however small it can be. The probability or a fact or event is the relationship between the number of favourable cases 'p' to this fact with the amount of possible cases, assuming that all the cases are equally possible. The way of establishing the probability is what is called calculation of probabilities. However, it is from these two main laws of statistics that are inferred those which are the cornerstone of the sampling method, such as:

- The law of statistic regularity, according to which a set of units, randomly taken from a N set are almost certain to possess the features of the bigger group.
- The inertia law of the large numbers, which is a corollary of the former, and it refers to the fact that in most phenomena, when a part varies in one direction, it is likely that an equal part of the same group varies in the opposite direction.
- The law of permanence of the small numbers, that is, if a sufficiently large sample is representative of the universe of study, a second sample of the same size must be similar to the first. If in the first sample few individuals with odd features are found, the same proportion is to be expected in the second sample.

Consequently, the qualities of a good sample for it to have technical-statistic validity can be summed-up in the following way:

- Being representative or a general reflection of the studied set or universe, reproducing the most exactly the features of said universe.
- That its size is statistically proportionate to the magnitude of the universe.
- That the sampling error is kept inside the limits regarded as admissible.

In the current work with users and mobile phones we have established a reduced universe of study, due to the fact that our methodology does not require analysis labs, because essentially we rely on the presence of a new professional known as communicability expert. Consequently, the study made with real users has been useful to verify some of the results obtained beforehand and in an autonomous way.

The task of determining a sample is included inside the different stages of research, consequently the sampling task (design and collection of the data that make up the sample of previous and latter tasks of what is known stages for the selection of the sample. These stages can be summed up in the following way:

- To keep always in mind the main and secondary goals of the research.
- To have the most possible available information about the set from which the sample will be taken.
- Allowed error of estimation.
- To count the available human resources and instruments.
- To examine the methods and techniques to be used in the research.
- To compile and analyze the data.
- To measure the representativeness of the sample.

Now the design of the sample and the sampling plan is an operation that demands special attention and training. Therefore, in many occasions the researchers need to count on specialists in sampling. The latter and former tasks are part of the general stages of the research but they must also be known by the statistician responsible for the selection of the sample. For instance, currently there are many works in some provinces of Northern Italy, such as can be Bergamo, where alleged works of educational research of the universe of secondary studies of the province sound the false alarm of the damage that the Web 2.0 does among teenagers. However, these studies lack scientific value because they do not follow the technical and theoretical patterns of the samples, like the stages for their selection, for instance. The falseness of the result is due to the fact that the sample does not represent the principles and main laws of descriptive statistics. Inside this context of stages in the selection of a sample, it is important to differentiate the types of it, the selection procedures and the representativeness. If one considers the structure and the selection procedures, one can distinguish two types of samples and inside each one of them there are several kinds [50]:

1) Random sampling, also known as probabilistic or randomly, is based on certain laws, and it is rigorously scientific. In our work method we are always based on this kind of sampling.
2) Non-random sampling, also known as empirical or erratic, which in contrast to the former does not have a statistic-mathematic basis, and may take two forms: intentional sample or circumstantial or without rule.

The random samples are obtained through procedures based on the law of large numbers and the calculation of probabilities, thus eliminating possible arbitrariness with a random determination. It takes for granted that the universe of study can be subdivided into units known as sample units. These sampling units may be natural, for instance, the users of multimedia mobile phones of a city; natural numbers, for instance; university students; or artificial units, such as can be the surface in meters or square kilometers that a university occupies. By claiming that a sample is randomly determined, one is stating that any of the units or elements that make up the set has the same possibilities of being included in the sample. It appears obvious that this procedure has higher chances of validity the more homogeneous is the set. Goode and Hatt [51] contend that a random sample is that which is obtained in such a way that from all the pertinent points of view the researcher does not have any reason to believe that it may cause any propensity or tendency. That is, the units of the universe have to be arraigned in such a way that the selection process yields an equal probability of selection to each and every one of the units that make up the universe of study. Although there are several procedures for the random sampling which give rise to different kinds of samples, in the current work we will focus on those which we frequently use in the set of techniques and/or methods of the communicability and the quality of the design of the interactive systems, including mobile phones.

- Simple random sampling: It is the foundation of any probabilistic sampling and consists in that all the elements have the same probability of being directly chosen as part of the sample. Two cases may turn up:

1. The sampling without replacement, used if the population is finite, in which all the samples of 'n' elements are equally probable.
2. The sampling with replacement, in which every selected element returns to the set or universe. These are then finite populations or universes.

- Stratified sampling: With the purpose of improving the representativeness of the sample: when certain characteristics of the set or universe of study are known one groups them in strata or categories, the sampling units which are homogeneous among themselves. Inside each one of the strata one proceeds to a random selection, that is, random simple. The distribution of the sample in every strata (technically denominated affixation) can be carried out through a random selection and in such a way that:

1. Each strata has a sample of the same size (uniform affixation).
2. Every strata is proportional to the number of elements of every strata (proportional affixation).
3. The sample of every strata is proportional to the number of elements and the standard deviation. In that case the optimal size of the sample for every strata has to be determined.

Finally, there are the non-random or empirical samples which are not based on a mathematic-statistical theory but which depend on the researcher's judgment. Very used in the regions of Northern Italy in the university educational context and the new technologies, especially when one tries to study wide territories of population, such as can be a whole province. In relation to the random samples, this method possesses advantages in regard to the cost-time equation, but it is more difficult to control the validity of the results. There are two modalities of non-random samples: the intentional sampling and the erratic, circumstantial and without rule sampling. The first one offers certain minimal guarantees in regard to the results because there is a person who chooses the sample intending it to be representative, doing it in agreement with his intention or opinion (it is the classical modus operandi of a star enunciator devoted to the research in the university context, for instance). In contrast, the second is the common denominator detected in the centers of university education, especially in the area of the Mediterranean, where numerous works of the usability evaluation are carried out by using the cases that are at hand or which have to be selected arbitrarily.

The sampling error is present even at the moment of using the best procedures since no sample can be an absolute guarantee of being an exact replica of the universe of study it represents [39] [47] [52]. That is, errors are unavoidable in any sample. This difference between the real universe and the sample of study is called sampling error. The important thing is to be able to determine the order or margin of the errors and their frequency inside the set since the latter allows us to establish the confidence interval inside which we are moving. Usually two kinds of mistake are distinguished: systematic or accidental. Systematic mistakes stem from several sources which, although they are alien to the sample, generate distortions. These distortions make the obtained results veer towards a direction in particular. Some examples of these systemic mistakes are the omission errors due to observation errors of the researcher, which is why they are usually called human factors; the insufficiency in data compiling, especially in the questionnaire technique (lack of responses), the inadequate replacements of the elements that make up the sample, that is, when it is not possible to have access to the correct elements and one resorts to those within reach, for instance. The second kind of error, called accidental or random mistake, is that in which the average values obtained from the sample will differ from the real average values of the universe. These differences are generated in the randomisation or in the measurement instruments. The sampling error depends on two factors:

1. The size of the sample: the greater the size of the sample, the lesser will be the error.
2. The dispersion or typical deviation of the sample: the greater the dispersion, the greater the error.

Consequently, the communicability evaluators must avoid to the best of the inability the systemic or random errors, starting with the personal human factors and the work team (evaluators and users of new technologies), if there was at the moment of carrying out the tests or remarks, for instance. The human factors in some places can severely damage the validity of a heuristic research, because those responsible are guided by the principles of what Saussure calls "parochialism" [39] [53]. That is, being against the interrelations among human beings beyond the local gridlock geographically speaking and even outright rejection of novel things, deriving from scientific progress. The notions and solutions deriving from computing, software engineering, usability engineering, human-computer interaction, ergonomics, etc. to eradicate that source of error and increase in the production costs of the interactive systems in the last two decades.

SOFTWARE AND HARDWARE EXPANSION FOR MULTIMEDIA MOBILE PHONE: NEW TECHNOLOGIES, USERS AND MARKETING

The mobile phones have developed so quickly in the last five years that the great majority of them currently possess a set of functions, similar or superior even to a Pocket PC of the early millennium. Conversely to what has happened in the evolution of the computers among the operative issues for Macintosh or Microsoft compatible with IBM PC, in the context of the mobile phone each hardware manufacturer presents his alternative in software for the functioning of their products. Obviously this is a great advantage for the advance of the potentialities of the device in the telecommunication issue, but it is also true that the adult public can't interact quickly with each one of these models without a previous experience. That is, there is no standard universal in those operative systems, for instance. Here is the reason why one of the design categories of the interactive system such as compatibility has a very important role.

One of the main components of compatibility is connectivity [39]. Great part of the economical and industrial activity regard the mobile phone as a device to be connected. In this sense connectivity has been boosted. Technologies such as Bluebooth or the infrared have lost importance with regard to the WiFi and 3D connectivity. Even the possibility of having access to the Internet from the mobile phone is becoming some necessary for a growing number of users. Consequently, in the middle of the current decade almost all mobile phones will need to be wired to the Internet. The appearance of the 2.0 Web is a determining factor in this phenomenon. Obviously, this factor can be either slowed down or boosted if free accessibility to information is maintained. However, from the sector of on-line information, such as the digital press, they have started to close down the free access to on-line contents. The user can only have access to the first page of the digital paper, for instance. However, the size of the screens of the mobile phones have not been convinced to supplant an E-book, for instance. Now there is no operative system that doesn't count with several alternatives to be able to connect with Facebook, Twitter, etc. We have gone from the massive communication era to the personal communicability era or in virtual communities, where the software and the hardware of the mobile phones is a determining factor in daily life [54] [55] [56].

Inside this new expansion age of communicability a variegated spectrum of applications for the mobile phone has emerged. In the same way as happened with the databases in 1980 and which has been maintained along the decades, the important thing is transportability and/or compatibility of these data or files as the software of the applications evolve [57]. As happened at that time to the databases, the present is a period of great activity in the design environment and the programming of applications in the mobile phones for millions of users across the planet. The origin of the current programming phenomenon is to be found in Apple's App Store. Since its conception, all the multimedia mobile phone makers have had their own services available. Some of them are works in progress, such as is the case of App Catalog of WebOS, others, in contrast, have a trajectory in the sector such as Android Market. Obviously, these examples like others, do not reach the market figures as for services of the Apple iPhones. The great advantage of this phenomenon is that so far the operating systems of the commercial mobile phones were regarded as a fenced-in system, that is, without enlargement possibilities. In contrast, now we can widen its functionality to carry out any kind of tasks, since the terminal itself. The updating or widening of functions will be made in an immediate way. For instance, we can get the street guide of a bordering country if we want to circulate through its highways and/or roads network. We can also update the videogames in the case that we have missed a train or a plane and we have to wait for the next one to leave towards our destination, etc. However, in the view of this great possibility of updating, it is always necessary to make a diachronic analysis of the applications as it is made in the multimedia interactive systems and the mobile phones are not an exception. The reason is that many applications for which the provider cashes in from the potential user are just adaptations to the already existing software. The problem lies in the interfaces of said applications because in some cases they have not been adapted to the new screens of the multimedia mobile phones. For instance, Windows Mobile needs to get updated, otherwise it may become obsolete from the communicability and usability point of view. Currently WebOS has been devised to be always wired to the net, to the extent that it has been conceived since its design stage to be always informed about the latest news and keeping the possibility that our data are accessible from any place in the world. That is, one is boosting the artificial life but reality on-line of the information stored in our mobile phone devices. We say artificial life because in some way the mobile phones will exchange data with the network in an automatic way, including the updating and real because in contrast to artificial life with artistic purposes, such as the fractal algorithms to generate tridimensional images, the data that we handle are vital to the telecommunications with commercial, industrial, educational purposes, etc. That is, the daily life of the human being in the economically developed societies.

HUMAN-COMPUTER INTERACTION AND INTERFACE DESIGN

In this daily life there are automatic cashpoints with tactile screens where some Southern Europe countries have the greatest network of international banking information such as in Spain. That is, information and banking operations have been within the reach of everyone since the 90s [14] [58] [59]. This technology already existed in some screens of the Hewlett Packard computers of the 80s, always with a commercial and/or financial purpose. Simultaneously, the first stands of tourist information were also created in the buses, trains, airports stations, etc, with the use of tactile screens. Later on, this way of interacting with the interactive systems has been reduced in size such as mobile computing screens like Palm, Tablet PC, E-book, etc. [60] [61] [62] [63] [64]. In these small size devices the tip of a pencil is used to activate the several functions of the programmes instead of the mouse, for instance. In the first models of the mobile phones keys were always used like in the keyboard of a calculator. However, from 2007 onwards we have entered the tactile revolution of the mobile phones like the iPhone, multimedia phone with Internet connection, tactile screen (multitactile technology) and a minimalist interface. It was from then on that almost all the mobile phone makers had to adapt their products to the new requirements of the users to have a tactile screen available.

Obviously, this also entailed the adaptation of the operating system to the new demands. In this speed of adaptation the communicability of many interfaces has been left aside to face the international demand of the new phone models which still maintained in their tactile screens the same old functionalities, shapes, topology of the interface components, etc. as in the classical phones with keyboards. In short, a kind of opportunism took place in the design of the interfaces, that is, to change everything so that nothing changed (gattopardism). To such an extent that the fast remodelling of the software to adapt it to the demands of the international market has prompted that we are in the face of interactive systems for mobile phones which although are handled with the touch, they still keep the same design as before. In an endless number of cases, these are hard to use and require expert users in computing. Consequently, the communicability and usability of these telephonic devices gets lost.

In the constant evolution of the new technologies and in a particular mobile microcomputing joined with telecommunications, the changes that take place daily, can make totally obsolete as a novelty a product that has been promoted for a week on digital television, for instance. The reason for this has to be looked in the hardware revolution that is always ahead of software. In the context of mobile phones, the revolutionary changes linked to software and hardware can derive both from the design of the interactive software or the ergonomic point of view, the appearance of new components to cut down even more the size of the phone devices, the incorporation of new materials to lengthen the load of the batteries or make more agreeable the vision of the information on the mobile phone screen, the incorporation of new devices or the adaptation of others which exist for constant connection, such as can be Bluetooth , etc. (see figures 7 and 8). In all these cases, you always have to differentiate the inventions from the technological discoveries. That is, that which didn't exist before in the context of the new technologies, for instance, or the new use of something which already existed. In this sense, mobile phones is an interesting field of intersection of inventions and discoveries.

In the history of the new technologies, as a rule, a consensus was usually generated among the different manufacturers of the new products to set up a series of quality standards, not only from the point of view of the product and/or services, but also of the potential users [65] [66]. Nevertheless, this modus operandi has all but disappeared from the mobile phone scene, where each manufacturer, Asian, European and North American has set up his/her own rules, trying to impose them to the rest of the potential users. Evidently, there are elements which are some kind of common denominator: keyboard, screen, batteries, servicing costs, coverage, etc. In principle, this common denominator has been easy to manage, both from the hardware and software point of view, for instance. In contrast, the cost factor of the service has served to slow down the democratic spread among the population. This factor deserves a detailed analysis, as it will be seen in the next sections, because it prevents in many cases the evolution of the population's quality of life through the new technologies.

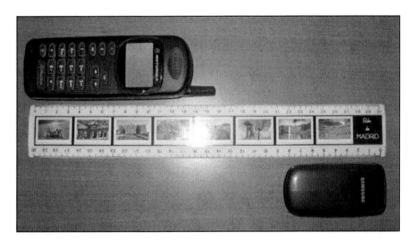

Figure 7. "Motorola" and old mobile phone: heavy and long –16 centimetres (no multimedia) and a new mobile phone "Samsung" E1150 (light and short but closed –8 centimetres).

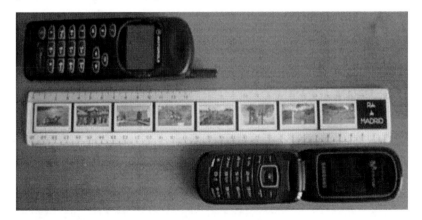

Figure 8. Once the Samsung model is open (16 centimetres), it has the same size as the Motorola. Because of ergonomic reasons, the users whose age surpasses 60 years prefer that or another similar model.

The operative system in the mobile phones has served to gain the potential users of the hardware. In this regard as usual the lines followed by Microsoft and its operative system Windows on one side, and Apple on the other side are worth mentioning. From the point of view of the design of the interface component it is easy to detect in the first operative systems how there is a transfer of an endless number of features (dropdown menus, scrolling bars, activation and applications shutting, etc.), similar to the Windows operative systems or those used in Apple. Consequently the features of the design of the interface, the communicability, the software and its updating (applications located in on-line virtual shops), the hardware and its potentialities in the user's mobility [5], among other main features lead to the trends of the multimedia mobile phones being concentrated on a few brands, mainly distributed across Asia, America and Europe.

ROCOCO OR MINIMALIST DESIGN FOR MOBILE INTERFACES

On the screen of the different models it can be seen how there is a prevailing style in regard to the amount of functionality of the icons present in the interface [67]. In the interface of multimedia mobiles and for space reasons we can find either few or many icons. That is, a minimalist style, in the first case, which sometimes implies speed in the performance once the user knows how to navigate through the icons and even the same icon can have several associated functions, unfolding a menu, for instance. In the opposite sense, we can come across screens where each icon has a given function and consequently the required number of icons is bigger and it may be that they do not unfold associated windows [16]. If there is communicability in the design, the main advantage in this case is the ease of use, especially for those adult users who do not have a usual or constant familiarity with mini screens of the mobile computer, for instance.

In the wide range of the operating systems that work in the mobile phones it can be seen how the dynamic and static means are present in computer devices aimed at the small-sized telecommunication. These operating systems allow us to carry out tasks such as recording high quality video, obtain high quality pictures to generate a high quality personal agenda, etc. In the current work we will focus on some of the following mobile phones with tactile screens that are next listed. Evidently many of these features will be outdated by the brands that are mentioned at the moment of reading of the current chapter, however their main features have established different differentiation lines which throughout time will have either joined or excluded each other. The models and their makers are (in brackets are the initials of the different models):

- Android 1.6 (HTC Tatoo) is an advanced model inside the Android range. Additionally, the low resources consumption of the operative system allows for a remarkable speed in the different applications which can be stored in it. The only drawback is the size of the screen, which is smaller in regard to the HTC-HD2 model.
- Blackberry OS 5 (Blackberry Store 2). The Canadian firm incorporates a tactile screen with a haptic technology which acts like a great button and which upon writing triggers a pleasant feeling to the user as if he/she were writing on a keyboard.

- iPhone OS 3.1 (Phone 3GS 16Gb). This embodies the evolution of Apple's hardware. It has a high RAM memory and a very fast processor.
- Symbian S60 V5 (Nokia 5800). It is a mid-range model with a high quantity of multimedia functions.
- WebOS (Palm Pre). Through the current model the firm Palm has released to the general public its new operational system. On it is a slide design keyboard is incorporated and the touchstone technology allows to reload its battery in a wireless way.
- Windows Mobile 6.5 (HTC HD 2), the HTC initials refer to the manufacturer in Taiwan, that is, High Tech Computer Corporation. It is a model with a high number of functionalities and with a 4,3 inches screen. It has a IGHz Snapdragon processor which allows to make work quickly the operative system it carries.

It has been seen once again how in the evolution of the software and the hardware applied to the new technologies in telecommunications, the designers, manufacturers and marketing work in an interdisciplinary way to obtain the highest profit in the least time. However, in the initial process of the design and conception of the new products and services in mobile phones it is important to consider communicability between the new devices and the potential users.

In the context of the early mobile phones and those of the first of the new millennium the following phone models were analyzed (users in the HCI lab): Siemens C45, Siemens A65, Motorola RAZR V3x, LG U81200, Samsung E1150, Samsung SGH-L170. Besides, we worked with Blackberry Storm, HTC HD2, HTC Tatoo, iPhone 3GS, Nokia 5800, and Palm Pre. In all these models there are several common denominators, such as the phone card, the connexions to the different peripherals (loudspeakers, keyboards, earphones, etc.), the batteries [8], etc. The energy consumption of this kind of phone is the Achilles heel of several models since they require to be connected several times every day to the electric network. The ideal solution today is that presented by WebOS, with the Touchstone technology. The multimedia functions require a high activities load for the processor, which entails a high energy consumption. It is the case of Internet navigation through Wifi or 3G which quickly empties the battery as also through the use of the GPS location systems. The mobile phones are replacing little by little the devices for such ends or the Palm PCs. This Achilles' heel has not evolved at the same speed as the rest of the hardware components. Equally, from the point of view of software a series of strategies have been activated to lengthen the duration of the battery loads as will be seen further on. Next the last generation multimedia mobile phones will be grouped. However, many of them do not emulate anymore the early models of the mobile phones. Consequently, for the adult population, some manufacturers have gone back to the early models, reducing considerably their size, thanks to the new materials for the hardware, and with accessible costs for most of the population, such as is the Samsung E1150, with a 65.000 colors screen.

SETS AND SUBSETS OF THE MOBILE PHONES IN THE COMMUNICABILITY ERA

A way to understand the state of the art of communicability inside the current technological context consists in presenting and breaking down the main characteristics of the mobile phones such as their main positive and negative aspects. These comparisons must take into account both the diachronic and synchronic of the technological evolution. Since the first monochromatic models of the 90s we have seen how the size of the original tube of the classical phone has been cut down until occupying the palm of a hand [61]. Then the colour screens have appeared the picture and video cameras have been incorporated, multimedia reproduction, ubiquitous data collection, etc. [15] [68] [69] [70].

Since 2010 the convergence of the tactile screens and the connection to the Internet we are in the face of a software generation which is trying to turn them into miniature computers. We can draw a dividing line in the set of the mobile phones generating two subsets. In the first are the new generation operative systems such as is the case of Android (Phone OS and WebOs). Whereas in the second is included that software that has been evolving along time, such as are Windows Mobile and Symbian.

In the first subset it can be seen how the design of the interface has a more modern appearance, with an intuitive navigation and the architecture to be wired to the net. At the same time we can see in the second group a whole operation of interface make-up, that is, the menus have been redesigned making more visible some of their components bearing in mind the greater size of their screens, for instance. Additionally, they have incorporated new functionalities. However, there are still menus of a classical or ancient style which still damage the communicability with the users. All of this happens inside the Windows Mobile. In new version 6.5 we can see how the marketing has bet on a make-up operation by calling it "Phone". Obviously, the possibility of moving across its screen without using a pencil is something that the young generations of users appreciate. However, the scrollbars and basic menus of the operative system still remain dodgy to our fingers and hard to press. Now being a veteran in the software sector has some advantages such as counting on a high number of software developers who constantly publish its applications. Many of these applications are free.

Here it is necessary to keep in mind that a free software does not entail a lower quality, but a way which some developers have to get known in the national or international market. It is not for nothing that we can find in the Internet free quality tools to carry out any kind of task with Windows Phone. For many of these tools it would be necessary to install the Net Computer Framework so that the programmers could develop efficient applications and without so many complications. The problem that this operating system currently has, but which derives from the manufacturer's commercial and marketing environment, that is, Microsoft, is the lack of a virtual shop to concentrate the sales of these applications, in the way that the Apple's App Store does (http://store.apple.com). These virtual shops allow millions of users to navigate through its catalogue and purchase those applications they are interested in having in their mobile. In contrast, Nokia is trying to generate a 100% renovated system aimed at Linux [17]. Between both subsets we can insert a third, which is the case of the Canadian Blackberrys, with a sober-looking interface but which reaches very well the communicability goals among a wide range of potential users of these devices in the adult population, for instance.

INTERNET NAVIGATION, APPLICATIONS AND E-COMMERCE MISCELLANEOUS

In regard to Internet navigation, all these models have the necessary components incorporated to implement this goal [71]. However, the current hurdle is the cost of the navigation through the mobile phones. As it happened in the dawn of the mobile phones age, it is necessary that the services offer diversifies in order to cheapen costs. Consequently, the use of the portable computer is a priority for many professionals who require hours of internet connexion. Inside the said context, Google, with the promotion of the Open Handset Alliance and the conception of Android as a free operative system has started to seek the free phone. So far both the hardware manufacturers and the mobile phone operators seemed to have agreed on making it difficult for customers to have free use of the purchased hardware. It suffices to remember the existing barriers inside the EU that forced to buy mobile phones in each country of the zone. Now, all the platforms have opened their gates to the developers so that they experiment with them through the different development kits, allowing a higher or lower freedom degree according to the case. The manufacturers know that the success or failure of their on-line trade is linked to the future of their operative systems. Consequently, the contribution of third firms or independent developers is a cornerstone of this new way of approaching the perspectives of the mobile technology in the coming years.

Google intends to develop a new operating system for mobile phones with an open code based on Linux. An aspect on which Apple and Google diverge is the software restrictions. Apple intends to keep a ring-fenced environment in the control over all the applications that work in the iPhone [1]. Additionally, it is necessary to pay certain costs for the SDK (Software Development Kit). Whereas the Android source code is freely available in the net within anyone's reach and the development environment can be obtained for free. In this way all the programmers who wish to can contribute to the development of the operative system. Android 1.6 is a fast, stable and 100% wired to the net system. In contrast to what happens with Microsoft phone, there is on-line the Android Market which allows to have a priori a global view of the applications that can be enclosed in the mobile phone. Evidently it does not reach the level of Apple's App Store. One of the problems of Android is the safety for it is the user who determines the permits for each application that he/she encloses in the mobile, for instance, Internet communication, sharing the stored personal information, the phonebook , etc. Although it is the user who makes the decision, it is possible that there are failings from the point of view of the safety of the information.

Since the 90s the name Palm has been linked to the pocket computer of millions of users in the whole world, especially for the handling of the personal agenda and the reading of e-mails, once downloaded from the computer [72]. At the start of 2000 it was also linked to the smartphones thanks to Palm OS. However, in the international commercial context Windows Mobile and iPhone have shifted it from the first spots in the preferences of the users in the mobile phones market. Yet it has been seen in the last few years how Palm is gaining ground inside the mobile phone sector with WebOS. On-line integration and the multitask applications management are the strengths of this system based on Linux. Upon navigating through the interface of the operative system we can see how the Cards interface (a windows in which every application will run on a different card and which allows to switch from the one to the other without wasting the work that has been made). On the desk you can switch from one card to the other and see them work, although they have a small size. The multitask system is very good, since one can see the changes in real time if in the middle of a process. Additionally, the implemented method is qualitatively high for the notifications of system events and of applications that are functioning on a background. In the case of the e-mail, they are located in the lower part of the screen, occupying a small space.

The devices presented so far that have WebOS incorporated are Palm Pre and the Pixi (gestures systems). Both present a zone outside the tactile recognition screen known as "gestures area" and in our days they make up an important part of the communicability and the usability of the operative system. The gestures are movements that the user makes with the finger in said area and on the screen to carry out certain operations, such as going backwards (it means to go back in the menus or getting out of the main menu from an application) through the shifting of the finger from left to right. Obviously here the cultural factor of reading is taken into account, since it differs with Eastern users. Not bearing in mind this aspect may be practical from the system's usability point of view, but it is little efficient from the point of view of communicability for a system conceived in Western culture and used by millions of people in China or Japan, just to mention two examples. Once the commands have been learned by the users, these are carried out in an almost natural way, and they remarkably speed up the interaction with the mobile phone, because you can do without the keys and the menus for some functions.

This operating system has been designed to be wired to the Internet 24 hours a day [4] to such an extent that several of its components do not work if we are not wired to the Internet (connecivity around the clock). That is, a great part of its applications have been thought to keep a constant data flow, carrying out periodical synchronization operations with Gmail or Google Calendar. Evidently, the cost factor of the constant phone connection has to be previously analyzed at the moment of plumping for such an operative system. Also, the energy consumption that makes the batteries to have a short duration and force the user to constantly connect to an electrical power source to reload them [8].

It is easy to see in the Internet how Palm hasn't put yet at the disposal of the developers of the applications to foster their offer a portal that groups them up as in the case of App Catalog. Obviously in this portal it is possible to detect those which are not yet 100% developed, and they have the Beta heading. Others in contrast present an interesting chance of incorporating additional functionalities to the phone, such as a language translator –a dictionary in nine languages), Pocket Mirror (synchronization with calendars, notes and contacts from Microsoft Outlook), Missing Sync (automatic synchronizer through the wireless net, the common programs data, for instance: Outlook, Windows Media Placer, iPhoto, etc.) [9]. In short, the interface is quick to answer and quite attractive visually. However, its applications are not sufficiently well categorized in the portal related to it.

iPhone like iBook or the iPad is always linked to software to set historic landmarks in the evolution of computing [1] [2]. The iPhone OS 3.1 is not an exception to this role due to the innovation of the software. Both the hardware and the software have set a pattern to be followed by the great majority of mobile phone makers. Although that star product was located because of its cost at the top of the pyramid of the democratic distribution of the new technologies when it comes to phones, now with the speed of the new operative system without the need of having available a powerful hardware, has moved it to the basis of the pyramid. The feeling of ease that you can enjoy in the menus is due to two main factors: the good design that Apple has traditionally developed along the decades and to the capacitating screen that these devices have to be handled through the fingers. However, operations like copying and sticking, something so usual in the alleged Latin creativity hasn't been possible until the third version of the operative system. What is really funny is that in the first version of this phone no multimedia messages, that is, MMS, could be sent.

Appe Store is a portal imitated by all the competitors of Apple. This virtual shop, since its opening until 2010 has surpassed two billions applications downloads. These are applications aimed at the basis of the technological pyramid where the secret lies in their cost, which is below 10 US dollars. Evidently, if we add to this the possibility of developing applications through the development kit, or iPhone SDK, the interest grows inside this pyramid of users which can be placed at its half, such as are programmers. The problem that has arisen is that many of these applications that are very interesting from the point of view of software engineering and especially in the context of the open software is that they harm the multinational's profit purposes. In this regard we can mention the possibility of carrying out free VoIP calls or the use of a terminal as a 3G modem for the desk computer. Without any doubt it is the portal where the user can find an endless and varied number of applications. Some of them are: the Evernote (it allows the user to keep in the form of notes a collection of textual or multimedia information that we want to remember, obtained from the Internet, for instance, and kept inside a cloud), Twitterrific (to send 140 characters in the communication), Quick Office (file editor in Word or Excel, visualize the contents of the files in Powerpoint, Pdf, Html, etc.), and Stanza (manager of the e-book library integrated to tuning).

The Subian S60 v5 can be found installed in other phones that do not belong to Nokia. This is another operating system that does not require a high hardware range for its correct functioning or obtaining the highest number of functions from it. In the new version there has been an attempt at leaving aside the use of the pencil as much as possible in order to adapt to the interaction through the fingers. Here is the reason why on the main screen we find big size icons for the main menus, but unfortunately the vertical scrolling bars are still used which prompts the use of a pointer. Another of the negative aspects is the inability to properly navigate through several applications which have been developed in Java in the previous versions, whose navigability was solved with virtual buttons on the screen. Although compatibility is insured in the installation of the applications developed in Java, it doesn't happen the same with the rest of the applications typical of S60, which can be installed but you can't navigate through them.

The quality of Nokia's virtual shop is lower than that of Apple's. Although there is the possibility that the user gets additional information to navigate through categories, ordination by popularity, description of applications, various comments, scores, etc. The information in this regard is not 100% reliable as in other cases of virtual shops, since it is made by the users of the virtual shop and the rest of members who make up the international community of mobile phone users, for instance. Nevertheless, it possesses an interesting community of application developers for this kind of hardware. Some of these applications which can be found in the virtual shop are: Fring (which focuses several communication services such as Skype, Facebook, Twitter, etc.), Battery Extender (it extends almost 30% the shelf life of a battery, disconnecting those features of the mobile phone that penalize it with a higher consumption, that is, GPS, Wifi, etc. in the moment in which they are not used). Easy Busy (it allows to avoid those incoming calls we do not want to answer), MobiSystems Office Suite 5 (an office automation solution that allows to create, see and edit Word files, Excel and Powerpoint, for instance). Qik (allows to broadcast that which is recorded in streaming, alive and direct), Handy W1 (to keep interesting access points when they are discovered and visualized through Google Maps).

In the current decade the Blackberry is a device which can be regarded as today's symbol of the high range of professionals inside the mobile phone context. The reason is due to the high reliability of the operative system and the safety of the data in the telecommunication process [17]. This latter issue is what makes it an ideal device for computer expert users. Now this doesn't mean that for its use you have to be an expert in computing, because precisely one of the main features is the ease of use. The menus are clear and simple with a minimalist interface and a truly practical conception design. With the incorporation of animations and 3D emulations of the keyboards in the interface the communicability of the new version is optimal. The possibility of having the QWERTY keyboard available. With the passing of time and the arrival of the new versions it was decided to eliminate this possibility by generating expectation among the users accustomed to it. The new strategy went through the Surepress system. That is, a capacitating screen which can be pressed and acts like a great button. Upon touching the screen this does not make a pulsation, but it would rather be the equivalent to move the pointer through it and it highlights the different objects of the interface. To activate it you have to press it by providing the user a feedback feeling.

In the Storm model they have incorporated 4 pressing points and it remarkably improves the user's experience. As the new versions of these models have appeared, it has been seen how the multimedia functions have been gradually inserted, for instance, in the video context, musical reproduction, etc. to the detriment of office or business automation. The aspects related to the Internet have also been improved. There are not many applications if compared to the Windows Mobile operative system, for instance. Currently there is the possibility of furnishing our mobile phone with plenty of videogames which use to the utmost the hardware potential. To such an extent that we have a variegated range of applications such as: Nice Office CRM (it is a very complete virtual office since from it you can manage the information in a very efficient way), Personal Assistant (it monitors the bank operations with the possibility of deactivating said application in the case of loss of the mobile phone), DriveSafely (it allows to visualize the messages received while driving and without the need of interacting with the phone), Call Blocker Professional (allows to hang up, send off the calls to the voicemail, keep them waiting, etc.), QuickLaunch (it admits the possibility of visualizing the interface and the menus or submenus of the most usual tasks, for instance), the Weatherbug (it anticipates weather forecast in the case of travelling, knowing beforehand the weather the user will find when reaching destination). To have access to the virtual shop to set an application known as Blackberry App World which sometimes doesn't work very well in some Blackberry models.

ASSESSMENT AND RESULTS

It is important to remember that in its origins usability engineering was not considered in a positive way, and even unlikely because of a series of methodology questions the possibility of defining quality attributes and starting from them to break them down in relation to the components to generate the metrics. Our metrics were aimed from the start at interactive communication and especially at quality, that is, communicability. These metrics have been examined and used in endless on-line and off-line interactive systems oriented at tourism, e-commerce, e-learning, etc. [20] [37] [38] [39] [43]. Besides, their constant use and positive verification with the tests carried out with other methodologies of usability engineering have allowed the establishment of a new professional called "analyst in communicability" [39]. For instance, in the evaluations made in Barcelona the evaluators of usability and communicability engineering previously to the evaluations made with the users, have carried out a series of planned tasks to evaluate the design of the multimedia mobile phones and be able to verify the tasks that had to be developed by the real users. All of them have come to trace the same errors in the interactive design of the multimedia mobile phones, for instance. The advantage of the proposed methodology is that we need no labs or special equipment, only a communicability analyst.

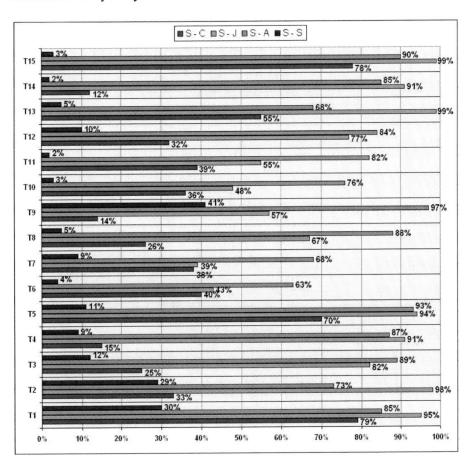

Figure 9. Results with Spanish (S) users (C = Children, J = Juniors, A = Adults, and S = Seniors).

The universe of study is made up by users of mobile phones (classical, multimedia and of last generation), hailing from two towns in Southern Europe: Barcelona, Spain (figure 9) and Bergamo, Italy (figure 10). In all the real universe amounts to 200 users, which have been invited to participate in the experiments as volunteers (all the food and transportation expenses were covered during the experiments). Those who have volunteered, were later classified in relation to age (children –4/11 years, Junior –12/17 years, Adult 18/64 years, Senior –65 or more years), sex, working experience, knowledge in computer science and English, the phone models they had available at the time, through a draw system, using a lottery approach. In this way were established the groups of users with whom work would be started. At the same time, the experts in heuristic evaluation in usability for multimedia and communicability systems started interact with the different models of mobile phones and set down a series of activities that had to be carried out by the users (in annex # 1 can be seen the table recording them). These activities were timed in minutes.

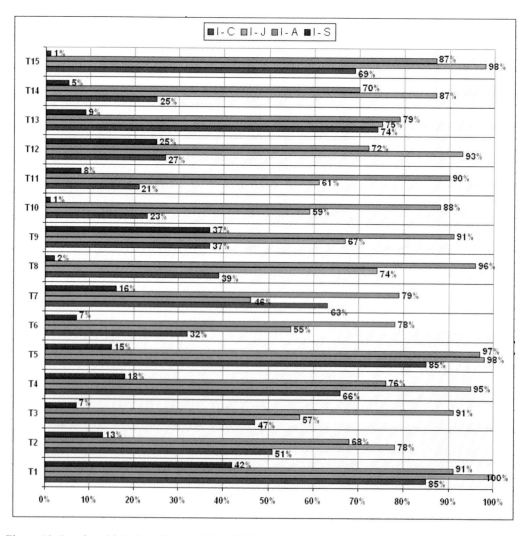

Figure 10. Results with Italian (I) users (C = Children, J = Juniors, A = Adults, and S = Seniors).

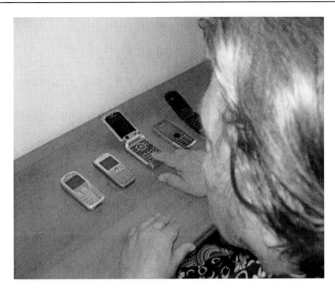

Figure 11. Senior user testing different models.

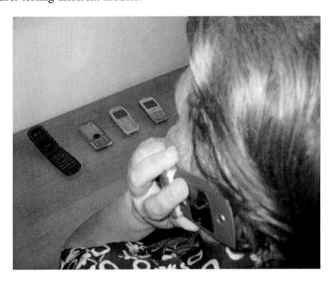

Figure 12. User analyzing the ergonomic aspect.

The users have been split into groups of 20, which were subdivided into four groups of five. Each one of them has been given the tasks to be carried out, for instance, sending a SMS, using the different models (figures 11 and 12). These operations had a maximum time to be carried out, and the result to be obtained was of the yes or no type. Some of the operations or tasks (in brackets the T indicates "task" and they are classified by the numbers in the figures 9 and 10 of the final results) to be carried out at the moment of the evaluation were: Having access to the phone with the secret number or PIN –Personal Identification Number (T1); Identifying the lost or made calls (T2); Making an enlarge –zoom in picture of a painting in the room (T3); Using the phone as a videocamera (T4); Reading or answering the received messages (T5); Writing down a phone number in the agenda (T6); Making arithmetic operations with the calculator (T7); Changing the volume of the sound of the calls (T8); Making international calls (T9); Tracing some streets in Google Maps (T10); Reading the headlines of the news in the front pages (T11); Recording messages (T12); Activating and deactivating call diversion (T13); Listening to music (T14); Playing with classical videogames (checkers, chess, cards, etc.) or in the net (T15). The evaluation techniques used were: observation, questionaires, videotaped sessions, and interview. However, direct observation has prevailed in the whole evaluating process.

The results obtained in the evaluation of the mobile phones with users inside the lab have made it possible to draw up a list of actions (vademecum) and conclusions which are next detailed:

- Usability must be split into several levels. One must start from very general problems which are common to all interactive systems and then aim at the specific problems of some given systems. That is, at a first level there are the mobile phones systems in general, at a second level the multimedia systems categories and at a third level the kinds of systems, for instance those aimed at communication or those aimed at work and communication such as a "Tablet PC" can be. At each level there must be a preparatory stage, a list of abstract tasks and a set of evaluation criteria.
- Upon evaluating a multimedia mobile phone system it is necessary to take into account first an inspection with an expert in heuristic evaluation of multimedia systems and in communicability. The results of the inspection are some application outlines, a list of potential problems and a set of abstract tasks. The tasks must be specific activities for users in the empiric evaluation.
- Empirical evaluation is more efficient in quality results and costs when there is a list of abstract tasks.
- The expert's inspection must be efficiently combined with empirical evaluation.
- Finally, there is the need to contact with the production of the evaluated system to advice the designers on the upcoming versions of the mobile phones, for instance.

It has been seen in the current work how the adult users whose age ranges about 60 years prefer first generation mobile phones, that is, phones that serve to communicate through the voice. Only 15% of them use the phones to send SMS. Consequently, the Motorola RAZR V3x and Samsung E1150 models are a real commercial finding aimed at this kind of users. The Italian and Spanish users with a greater purchasing power and whose age ranges between 16 and 40 years prefer the latest technological breakthroughs from the point of view of mobile phones. However, it is important to stress that only 8% among them know or at least have tried to use all the possible functions from the multimedia mobile phone. Among the young knowing such functions means to establish a virtual community. Adult people used to resort to the Internet or the salesmen or the assistance service about the functions of the phone. Currently, among children whose age ranges between 7 and 10, they start to use the mobile phone as a means of communication for the location of adult people (relatives, friends and schoolmates of their age). Among teenagers the mobile phone is used in 93% of cases to send SMS among friends and/or classmates in high school with occasional calls between parents, grandparents, siblings, etc., videogames, and currently there is the trend to trying to navigate from the multimedia mobile phone to have access to on-line videogames, for instance.

Figure 13. Senior user talking by the phone after having chosen the model that best adjusts to her daily needs.

CONCLUSION AND FUTURE WORKS

In the communicability, usability and usefulness triad a spate of difficulties has been seen which persists among the mobile phone users in the European countries of the Mediterranean basin. The presented methodology has pointed out quickly which are the errors of this triad. Said errors may damage the current era of communicability expansion we are living in. The quality attributes and their corresponding metrics must be constantly updated to adapt to the daily evolution of this sector of the new technologies, where the hardware goes ahead of the software. Consequently, it is important that the operating systems and the applications which allow the functioning of these devices take into account the different kinds of potential users, distributed all across the planet. The presented little vademecum is a first guide in this sense, since in Southern Europe the multimedia mobile phones may replace by the end of the current decade the portable computers, especially with the mobile computer, among the young users who constantly interact with computers and videogames. Secondly, for this group of users there is the TDTV and the cinema. In contrast, the adult public will keep on interacting with contents on the bigger-sized screens, that is, the PC, the portable computer, the TDTV and to a lesser extent the Tablet PC, the e-book and the PDA, for instance.

In the future we will aim our research at the mobile computer which uses the human body for its interaction operation with the interfaces that appear in the arms and hands of the users, for instance. We think that this will be the field towards which will derive the current multimedia mobile phones in the watch format. These are multifunctional watches whose size and shape can take up several of the current functions of the sensors of the mobile computer muscles, which currently are adjusted in the upper arm, under the armpits. It is based on EMG (Electromyography) senses this muscle activity by measuring the electrical potential between a ground electrode and a sensor electrode. This is a natural environment of the current era which is known as the communicability expansion era, where the users, through micro computing and telecommunications mobile devices interact with the interfaces which are projected from those EMG devices, to which can be linked the dynamic and static means of the multimedia interactive systems.

ACKNOWLEDGMENTS

The author would like to thank Emma Nicol (University of Strathclyde), Maria, Carlos and Miguel Cipolla-Ficarra (Ainci and Alaipo) for their helps.

ANNEX #1.

TABLE OF HEURISTIC EVALUATION FOR MOBILE PHONE

Age	Sex	Working Experience	Knowledge in Computer Science	English Knowledge

Task	Activities	Time (minutes)	Results Yes	Results No	Categoria Design
1	Having access to the phone with the secret number or PIN –Personal Identification Number				
2	Identifying the lost or made calls				
3	Making an enlarge –zoom in picture of a painting in the room				
4	Using the phone as a videocamera				
5	Reading or answering the received messages				
6	Writing down a phone number in the agenda				
7	Making arithmetic operations with the calculator				
8	Changing the volume of the sound of the calls				
9	Making international calls				
10	Tracing some streets in Google Maps				
11	Reading the headlines of the news in the front pages				
12	Recording messages				
13	Activating and deactivating call diversion				
14	Listening to music				
15	Playing with classical videogames (checkers, chess, cards, etc.) or in the net				

Design Categories: Layout (L), Content (C), Navigation (N), Panchronic (P), Structure (S), and Conection (Co)

REFERENCES

[1] Bondo, J. Barnard, D., Burcaw, D. and Novikoff, T. (2009). *iPhone User Interface Design Projects*. Apress: New York

[2] Macedoni, M. (2007). iPhones Target the Tech Elite. *IEEE Computer*, 40 (8), 94-95

[3] Tsai, F. et al. (2010). Introduction to Mobile Information Retrieval. *IEEE Intelligent Systems,* 25 (1), 11-15

[4] Smith, I. (2005). Social-Mobile Applications. *IEEE Computer*, 38 (4) 84-85.

[5] Faruque, S. (1996). *Cellular Mobile Systems Engineering*. Artech House Publishers: Boston

[6] Ramacher, U. (2007). Software-Defined Radio Prospects for Multistandard Mobile Phone. *IEEE Computer*, 40 (10), 62-69

[7] Seyff, N., Grunbacher, P. and Maiden, N. (2006). Take Your Mobile Device Out from behind the Requirements Desk. *IEEE Software*, 23 (4), 16-18

[8] Jacoby, G., Marchany, R. and Davis, N. (2006). How Mobile Host Batteries Can Improve Network Security. *IEEE Security and Privacy*, 4 (5) 40-49

[9] Reponen, E., Huuskonen, P. and Mihalic, K. (2006). Mobile Video Recording in Context. *Interactions,* 13 (4), 28-30

[10] Rao, B. and Minakakis, L. (2003). Evolution of Mobile Location-based Services. *Communications of the ACM*, 46 (12), 61-65

[11] Wang, A. (2006). The shazam music recognition service. *Communications of the ACM,* 49 (8), 44-48.

[12] Dimitriadis, C. (2007). Improving Mobile Core Network Security with Honeynets. *IEEE Security and Privacy*, 5 (4), 40-47.

[13] Balan, R. and Ramasubbu, N. (2009). The Digital Wallet: Opportunities and Prototypes. *IEEE Computer*, 42 (4), 110-102

[14] Herzberg, A. (2003). Payments and Banking with Mobile Personal Devices. Communications of the ACM, 46 (5), 53-58

[15] Giguette, R. (2006). Building Objects Out of Plato: Applying Philosophy, Symbolism, and Analogy to Software Design. *Communications of the ACM,* 49 (10), 66-71.

[16] Ozkaya, I. et al. (2008). Making Practical Use of Quality Attribute Information. *IEEE Software*, 25 (2), 25-33.

[17] Balasubramaniam, S. et al. (2008). Situated Software: Concepts, Motivation, Technology, and the Future. *IEEE Software*, 25 (6), 50-55.

[18] Toyama, K. and Dias, M. (2008). Information and Communication Technologies for Development. *IEEE Computer*, 41 (8), 22-25

[19] Mashey, J. (2009). The Long Road to 64 Bits. *Communications of the ACM*, 52 (1), 45-53

[20] Edvardsen, F. and Kulle, H. (2010). *Educational Games: Design, Learning and Applications*. Nova Science Publishers: New York

[21] Isaacs, E., Walendowski, A. and Ranganathan, D. (2002). Mobile Instant Messaging trhough Hubbub. *Communications of the ACM,* 45 (9), 68-72

[22] Lindholm, C. and Keinonen, T. (2003) *Mobile Usability: How Nokia Changed the Face of the Mobile Phone*. McGraw Hill: New York.

[23] Resnick, M. et al. (2009). Scratch: Programming for All. *Communications of the ACM*, 52 (11), 60-67.
[24] Biehl, M. (2007). Success Factors for Implementing Global Information Systems. *Communications of the ACM*, 50 (1) 53-58
[25] Ladner, S. (2010). The Essence of Interaction Design Research: A Call for Consistency. *Interactions*, 17 (2), 48-51
[26] Wilson, J. (2000). Toward Things That Think for the Next Millennium. IEEE Computer, 33 (1), 72-76
[27] Cipolla-Ficarra, F. (2010*). Persuasion On-Line and Communicability: The Destruction of Credibility in the Virtual Community and Cognitive Models*. Nova Science Publishers: New York
[28] Berners-Lee, T. (1996). WWW: Past, Present, and Future. *IEEE Computer*, 29 (10), 69-77
[29] Trowbridge, C. (2010). Marconi: Father of Wireless, Grandfather of Radio, Great-Grandfather of the Cell Phone, The Story of the Race to Control *Long-Distance Wireless*. BookSurge: New York
[30] Miller, F., Vandome, A. and McBrewster, J. ((2009). *Guglielmo Marconi*. Alphascript Publishing: Beau Bassin
[31] Ling, R. (2004). *The Mobile Connection: The Cell Phone's Impact on Society* (Interactive Technologies). Morgan Kauffman: San Francisco
[32] Green, J. (1993*). The Get Smart Handbook*. Macmillan Publishing: New York
[33] Nielsen, J. (1993). *Usability Engineering*. Academic Press: London
[34] Brown, J., Shipman, B. and Vetter, R. (2007). SMS: The Short Message Service. *IEEE Computer*, 40 (12), 106-110.
[35] Homquist, L. (2007). *Mobile 2.0 Interactions*, 14 (2), 46-47.
[36] Cipolla-Ficarra, F. MEHEM: A Methodology for. Heuristic Evaluation in Multimedia. *In Proceedings of the DMS'99. KSI: Aizu*, 89-96.
[37] Cipolla-Ficarra, F. *Communication Evaluation in Multimedia: Metrics and Methodology. Universal Access in HCI*. In Stephanidis, C. (Ed.). LEA:London 567-571.
[38] Cipolla-Ficarra, F. (1997). *Method and Techniques for the Evaluation of Multimedia Applications. HCI International '97*. Elsevier: San Francisco, 635-638.
[39] Cipolla-Ficarra, F. (2010). Quality and Communicability for Interactive Hypermedia Systems: Concepts and Practices for Design. *IGI Global Editions*: Hershey
[40] Cipolla-Ficarra, F. (1999). Evaluation Heuristic of the Richness. *International Conference on Information Systems Analysis and Synthesis*. ISAS: Orlando, 23-30.
[41] Miller, F., McBrewster, J. and Vandome, A. (2009). *Augmented Reality*. Alphascript Publishing: Beau Bassin
[42] Antoniac, P. (2008). *New User Interfaces for Mobile Devices Using Augmented Reality*. VDM Verlag: Saarbrücken.
[43] Cipolla-Ficarra, F. (1997). Evaluation of Multimedia Components. In Wemer, A. (Ed.), International Conference on Multimedia Computing and Systems. Ottawa, Canada: *IEEE Computer*, 557-564.
[44] Muller, M., Matheson, L., Page, C., and Gallup, R. (1998). *Participatory Heuristic Evaluation. Interactions*, 5 (5), 13-18

[45] Kuurniawan, S. (2007). Mobile Phone Design for Older Persons. *Interactions*, 14 (4), 24-25.
[46] Bunge, M. (1981). *The science: your method and your philosophy.* Siglo XXI: Buenos Aires
[47] Ander-egg, E. (1986*). Techniques of Social Investigation. Hvmanitas*: Buenos Aires
[48] Berg, B. (2001). *Qualitative Research Methods for the Social Sciences.* Allyn and Bacon: Boston
[49] Yule, Y. and Kendall, M. (1964). Introducción a la estadística Matemática. Aguilar: Madrid
[50] Canavos, G. (1988*). Applied Probablity and Statistical Methods.* McGraw Hill: Cambridge
[51] Goode, W. and Hatt, P. (1972). *Métodos de investigación social.* Trillas: México
[52] Hand, D. (2008). *Statistics: A Very Short Introduction.* Oxford University Press: Oxford
[53] Saussure, F. (1990). *Course in General Linguistics.* McGraw Hill: New York
[54] Smith, I. (2005). Social-Mobile Applications. *IEEE Computer*, 38 (4), 84-85.
[55] Halvey, M., Keane, M. and Smyth, B. (2006). Mobile Web Surfing is the SAME as Web Surfing. *Communications of ACM,* 49 (3), 76-81.
[56] Macedonia, M. (2005). Power from the Edge. *IEEE Computer*, 38 (12), 123-127.
[57] Ambler, S. (2007). Test-Driven Development of Relational Databases. *IEEE Software,* 24 (3), 37-43.
[58] Christou, I., Ponis, S. and Palaiologou, E. (2010). Using the Agile Unified Process in Banking. *IEEE Software*, 27 (3), 72-79.
[59] Holland, C. and Westwood, J. (2001). Product-Market and Technology: Strategies in Banking. *Communications of ACM*, 44 (6), 53-57
[60] Yoshioka, A. et al. (2006). Point, Push, Pull: The FAU Interface. *Interactions*, 13 (4), 40-41.
[61] Maiden, N. et al. (2007). Determining Stakeholder Needs in the Workplace: How Mobile Technologies Can Help. *IEEE Software*, 24 (2), 46-52.
[62] Prey, J. and Weaver, A. (2007). Tablet PC Technology the Next Generation. *IEEE Computer*, 40 (9), 32-33.
[63] Schneiderman, B. (2002). Understanding Human Activities and Relationships: An Excerpt from Leonardo's Laptop. *Interactions*, 9 (5), 40-53
[64] Kimel, J. and Lundell, J. (2007). Exploring the Nuances of Murphy's Law Long-term Deployments of Pervasive Technology into the Homes of Older Adults. *Interactions*, 14 (4), 38-41.
[65] Hinman, R. (2009). 90 Mobiles in 90 Days: A Celebration of Ideas for Mobile User Experience. *Interactions*, 16 (1)10-13.
[66] Kangas, E. and Kinnunen, T. (2005). Applying User-Centered Design to Mobile Application Development. *Communications of the ACM*, 48 (7), 55-59.
[67] Marsden, G. (2008). Toward Empowered Design. *IEEE Computer*, 41 (6) 42-46.
[68] Yi, J. (2010). User-Research-Driven Mobile User Interface Innovation: A Success Story from Seoul. *Interactions*, 17 (1), 48-51.
[69] Shilton, K. (2009). Four Billon Little Brothers? Privacy, mobile phones, and ubiquitous data collection. *Communications of the ACM,* 52 (11), 48-53.

[70] Yap, K. et al. (2010). A Comparative Study of Mobile-Based Landmark Recognition Techniques. *IEEE Intelligent Systems*, 25 (1), 48-57.
[71] Cao, J. et al. (2007). Data Consistency for Cooperative Caching in Mobile Environments. *IEEE Computer*, 40 (4), 60-66.
[72] Anokwa, Y. et al. (2009). Open Source Data Collection in the Developing World. *IEEE Computer*, 42 (10), 97-99.

Chapter 3

INCREASED GENETIC DAMAGE DUE TO MOBILE TELEPHONE RADIATIONS

A. S. Yadav[*1], *Manoj Kumar Sharma*[2] *and Shweta Yadav*[1]

[1]Department of Zoology, Kurukshetra University, Kurukshetra-136119, India
[2]Department of Zoology, Punjab Agriculture University, Ludhiana-141004, India

ABSTRACT

Mobile telephony is becoming increasingly integrated into everyday life. Recently mobile communication has experienced a vast expansion all over the world. With the advent of newer and newer technologies, particularly in the telecommunication, has made humans highly vulnerable to both ionizing and non-ionizing electromagnetic radiations (EMR) along with the natural background radiations. Electromagnetic fields now blanket the earth with a huge network of systems which emit these radiations in to the environment. The health risk issues related to the application of telecommunication systems became a worldwide concern. International Agency for Research on Cancer (IARC) has classified low frequency electromagnetic field as a possible carcinogen. New concept of how radiations interact with biological system is bound to suggest new approaches and new avenues for development of novel radio sensitizers for tumour treatment. Exposure of living cells to EMR activates genetic cascade of signaling events leading to cellular damage, primarily through a spectrum of lesions induced in DNA. This damage to living cells by radiations takes place at molecular level and can induce genetic instability. A number of biological effects induced by man-made Electromagnetic fields and radiations of different frequencies, including digital mobile telephony have already been documented by many research groups. During the present investigation 250 individuals, irrespective of age/sex/caste have been taken. Subjects exposed to mobile phone radiations (110) and healthy controls (140) matched with respect to age, sex and socio-economic status comprised the material. During the study no incidence of cancer or other diseases in persons exposed to EMR through mobile telephones was found. Comet Assay and Micronucleus (MN) Assay were used as the biomarkers of choice for evaluating genotoxicity. The mean frequency of micronucleated cells (MNC), total micronuclei (TMN) and binucleated cells (BN) showed an increase and the difference

[*]Email - abyzkuk@gmail.com, +919416173289 (Mobile)

was found to be statistically significant with respect to controls. Similarly, statistically significant difference was observed in the entire comet parameters between exposed and control subjects, thereby indicating DNA damage in the form of comet length, tail length, percent DNA in tail and Olive tail movement. The subjects exposed to mobile phone radiations showed a higher degree of genetic damage in DNA as compared to the control group. A positive correlation was observed between the duration of exposure and the extent of genetic damage in the mobile phone user group.

1. INTRODUCTION

Among the increasing environment related health issues and potential health hazards emerging from new, modified and combined exposure to various environmental factors, exposure to electromagnetic fields (EMF) at all frequencies and modulations represents one of the most common and fast growing environmental concerns. The certain way of interpreting potential hazards as risks is a reflection of ideological, political, ethical and moral values of the society. For the purpose of primary prevention, hazards identification is a first step (Vainio and Weiderpass, 2005).

The advancement of technology over the past 50 years in the area of biomedicine, telecommunication and industry has made humans highly vulnerable to the electromagnetic radiations (both ionizing as well as non-ionizing) along with the ever existing natural background radiations. Background level in urban, domestic and industrial environments have increased exponentially with the advent of electricity and electromagnetic (EM) communication systems. The use of new technologies, such as mobile telecommunications, is rapidly expanding in both developed and developing countries. Thus, health risk issues related to the application of these technologies have become worldwide concern (Groves et al., 2002; Ahlbom et al., 2004; Van-Deventer et al., 2005).

World Health Organization has ranked RF EMF the 'high priority' research field for assessing the health risk of RF EMF (WHO, 2006). Understanding the health impact of electromagnetic fields falls within the mandate of the WHO in the area of environmental health, which aims to help member states achieve safe, sustainable and health-enhancing human environments, protected from biological, chemical and physical hazards (Van-Deventer et al., 2005).

Mobile telephones now used widely, has been accompanied by an upsurge in public and media concern about the possible hazards of this new technology, and specifically of low radio frequency exposure. Despite the ubiquity of new technologies using RF's, little is known about population exposure from RF source and even less about the relative importance of different sources (Ahlbom et al., 2004).

During the past few decades, radiation research has developed into specialized sub-disciplines, from basic physics and chemistry to tumour biology and experimental radiation therapy (Dorr et al., 2000). Although the radiobiological effects are extensively investigated for X-ray and gamma-ray, little work has been directed towards low frequency non-ionizing radiation involved in mobile telecommunication devices. We do not know much about the health hazards of cellular telephones that are rapidly gaining popularity especially in developing countries like India. Given the immense number of users of mobile phones, even a small adverse effect on health could have a major public health implication (WHO, 2000).

The form of radiation consists of energetic particles and electromagnetic waves. However, low frequency fields are of very low energy and are not sufficient to alter DNA structure directly and cause genetic injury (Juutilainen and Liimatainen, 1986; Rosenthal and Obe, 1989).

Relevant genetic parameters for human biomonitoring studies include whether the exposure results in the induction of DNA damage, accumulation of damage with repeated exposure to elucidate the kinetics of the recovery process, which includes DNA repair and tissue regeneration (Smith et al., 2008).

Since RFs are invisible and imperceptible, individuals cannot directly report on their exposure, and therefore the quality of exposure assessment needs particularly careful consideration when interpreting epidemiological studies. In past a number of biomarkers have been used in environmental carcinogenesis research in the pretext of improving the reliability of the exposure assessment and providing tools for early detection of disease related changes and their association with environmental and genetic factors (Bonassi et al., 2006). Identification and validation of early-phase radiation biomarkers are needed to provide enhancement in biological dosimetry capability to assess individuals suspected of exposure to electromagnetic radiations (Blakely et al., 2005).

A wide range of methods are presently used for the detection of early biological effects of DNA-damaging agents in environmental and occupational settings. These include well-established biomarkers for chromosome damage measured by CA (chromosomal aberration) and SCE (sister chromatid exchange). However, both methods are laborious and time consuming, on the other hand, Comet assay and micronucleus assay are simple, rapid, inexpensive, and require little biological material.

The single cell gel electrophoresis (SCGE) or comet assay is an advanced visual and sensitive technique for the detection of DNA damage of individual cells in interphase (Singh et al., 1988). It is able to detect single strand breaks (SSB) and /or alkali-labile sites at the single cell level (Singh et al., 1988, 1994; Tice et al., 1991). It is based on the principle of quantifying the amount of denatured DNA fragments migrating out of the cell nuclei during electrophoresis. In recent years, this method has been extensively utilized for studies on DNA damage and repair, radiation biology, genetic toxicology and apoptosis (Singh et al., 1988, 1994; Tice et al., 1990, 1991; Tice, 1995; Collins et al., 1993; McKelvey et al., 1993; Fairbairn et al., 1995; Garaj-Vrhovac and Kopjar, 1998; Kopjar and Garaj-Vrhovac, 2001; Dillon et al., 2004; Pitozzi et al., 2006; Collins et al., 2003, 2008).

The micronucleus (MN) assay is a multi-endpoint test of genotoxic responses to clastogens (Fenech and Morley, 1986; Antoccia et al., 1991; Pinto et al., 2000; Fenech and Crott, 2002; Norppa and Falck, 2003; Kirsch-Volder et al., 2006). Micronucleus test is used by the academics, industry and contract laboratory organizations for internal hazards identification and compound prioritization as an alternative/replacement of the *in vitro* chromosome aberration test (CAT) as it offers significant advantages over the CAT (Corvi et al., 2008). The frequency of micronucleus is extensively used as a biomarker of genomic instability, genotoxic exposure and early biological effect in human biomonitoring studies (Norppa and Falck, 2003; Joksic et al., 2004; Yesilada et al., 2006). The test allows the detection of both clastogens and aneugens and it can simultaneously detect mitotic delay, apoptosis, chromosome breakage, chromosome loss and non-disjunction (Parry and Sorrs, 1993; Kirsch-Volder, 1997). It is used as an indicator of genotoxic exposition since it is associated with aberrant mitoses and consists of acentric chromosomes, chromatid fragments or chromosome aberrations (Belien et al., 1995; Ramirez and Saldanha, 2002; Thomas et al., 2008).

In populations exposed to genotoxic agents the comet assay mainly reflects recent exposure to clastogens and the micronucleus assay is rather used to assay genetic damage rates and cancer risk.

As societies develop, one can expect that the increasing demand and use of certain technologies will lead to greater exposure to EMF fields and subsequent need to study the potential impact on human health and the environment is expected. New concept of how radiation interacts with biological system is bound to suggest new approaches to protection involving probiotics, and new avenues for development of novel radio sensitizers for tumour therapy. The identification of mechanisms other than DNA damage as being involved in radiation response means that new intervention points for protection or for therapy are possible.

A major area of concern is the possibility that RF radiation from mobile phones is carcinogenic. Carcinogenesis at the cellular level is a multi stage process and if RF exposure is involved it would have an effect on one or more of these cellular stages. Most of the known carcinogens, but not all, are genotoxic, i.e. they cause DNA or chromosomal damage (NRPB, 2003). Therefore, if RF radiation were carcinogenic it could possibly also have genotoxic effects on cells. Many studies are devoted to testing for genotoxicity and use a range of *in vitro* tests to investigate this possibility (Juutilainen and Liimatainen, 1986; Rosenthal and Obe, 1989; Tynes et al., 1996; Lindahl and Wood, 1999; Morgan et al., 2000; Somosy, 2000; Groves et al., 2002; Ahlbom et al., 2004; Van-Deventer et al. 2005; Blakely et al., 2005; Zeni et al., 2005; Scarfi et al., 2006; Speit et al., 2007; Agarwal et al., 2008; Karinen et al., 2008). Several studies also test the possibility that RF radiation acts synergistically with other known carcinogenic agents to enhance or promote their effect (Maes et al., 1996; Ramirez and Saldanha, 2002). In India scanty information is available in literature with respect to this area (Gadhia, 1998; Gadhia et al., 2003; Ingole and Ghosh, 2006; Agarwal et al., 2008; Yadav and Sharma, 2008).

The main aim of the present study was to investigate the genotoxic effects of low frequency non-ionizing radiations, emitted by mobile phones on humans.

2. BIOLOGICAL EFFECTS OF ELECTROMAGNETIC RADIATIONS

The biological effect of radiation exposure can be classified as either stochastic (random) or non-stochastic (deterministic) (Davis, 2006). International Agency for Research on Cancer (IARC) has classified low frequency electromagnetic field as possible carcinogen- a categorization that necessarily implies that low electromagnetic field may promote DNA damage and hence may be genotoxic (IARC, 2002).

At the cellular level the radiation response may be manifested in irreversible changes, such as mutations, malignant transformation, development of abnormal cell-forms and the death of cells or as minor reversible structural and functional alteration of biological systems. Both the irreversible and reversible damage of cells are manifested at the sub-cellular level as structural and functional changes in various cell organelles (Somosy, 2000).Exposure of living cells to electromagnetic radiations activate genetic cascade of signaling events leading to cellular damage primarily through a spectrum of lesions induced in DNA. Recognition of these damages leads to the activation of specific proteins that in turn activate repair, cell cycle checkpoints, and apoptosis (Lavin et al., 1999). The damage to a living cell by radiation takes place at molecular level (Glasstone and Jordan, 1981) and can also induce genetic instability (Kadihim et al., 1995). The cellular response to various form of radiation, including ionizing-, UV- and non-ionizing irradiation or exposure to low frequency electromagnetic fields was manifested as irreversible and reversible alteration in the structural and functional changes to cells and cell organelles. In addition to morphological signs related to cell death, there were several irreversible alterations in the structure of the genetic material of the cells (Somosy, 2000).

Mobile telephony is becoming increasingly integrated into our everyday life. In the past few years, mobile communication systems have experienced wide and rapidly growing use all over the world. For this reason, there was a great concern about the effects of electromagnetic exposure of the cellular telephone frequency range on human population and in particular young generation (Hermann and Hossmann, 1997). Increasing use of mobile telephony in our society has brought focus on the potential for radio-frequency (microwave) electromagnetic radiation to instigate biological stress responses, in association with potentially detrimental effects of this to human health (Hardell et al., 2003). Area of particular interest were the possible long-term risk to health from occupational exposure to intense static magnetic fields, electromagnetic radiation (ionizing and non-ionizing), potential development effects and the threshold for acute risks to health (Van-Deventer et al., 2005).

A number of biological effects induced by man-made electromagnetic fields (EMFs) and radiations of different frequencies including digital mobile telephony have already been documented by many research groups (Guelman et al., 2001). These include changes in the cell proliferation rate , intercellular ionic concentrations, change in the synthesis rate of different biomolecules, changes in the reproductive capacity of animals, change in gene expression, DNA damage, cell death and these can lead to cancer (Lai and Singh, 1995, 1996, 1997, 2004; Verschaeve and Maes, 1998; Somosy, 2000; Ahlbom et al., 2004; Aitken et al., 2005; Ahmed, 2005; Remondini et al., 2006; Panagopoulos et al., 2004, 2007; Panagopoulos, 2008; ICNIRP, 2008; Jiang et al., 2008).

Earlier it was assumed that exposure to low-frequency, low-intensity magnetic field could not pose a health hazard but this view was challenged by Wertheimer and Leeper (1979) who studied the incidences of cancer in children with respect to the electromagnetic radiations of low-frequency. This was supported by another study (Savitz et al., 1988) reporting that children living in homes with potentially high magnetic fields had greater incidences of childhood leukemia than children living in homes expected to have lower magnetic field exposures. They also studied the DNA strand breaks in irradiated and control cell with DNA unwinding techniques and were able to detect effects of radiations down to a dose of 1 rad., corresponding to 10-20 DNA single-strand breaks per cell.

Garaj-Vrhovac et al., (1992) reported an induction of chromosomal damage by non-ionizing radiation. A significant increase in MN frequency was reported. It indicated that exposure to non-ionizing radiations in the range of microwaves was able to cause an increase of lymphocyte MN frequencies along with effect to cell cycle progression.

Balzano et al. (1995) measured the energy from a mobile phone radiation in a study involving human models and reported that energy accumulated during the process is mainly deposited in the cheek anterior to the ear with lesser level elsewhere. The deposited energy was approximately 1 W/Kg in the cheek region mentioned above.

Hocking (1998) published a preliminary report of symptoms associated with mobile phone use based on the survey and unpleasant sensations such as burning feeling or dull ache occurring in the temporal, occipital or auricular areas were reported.

The first nationwide cohort study in Denmark on incidences of cancer with respect to cellular telephones (Johansen et al., 2001) provide no support for an association between use of these telephones and risk of brain cancer, leukemia, salivary gland cancer, or other site specific cancers. McMillan et al. (2001) proposed the use of DNA double-strand break (DSB) quantification in radiotherapy as DSBs had been considered to be the most important type of lesion for the cytotoxic effects of radiations.

Weinberger and Richter (2002) hypothesized that the human head can serve as a resonator for the electromagnetic radiation emitted by the cellular telephone, absorbing much of the energy specifically from the wavelength used by the cellular telephones i.e. 900 MHz for analog and 1800 MHz for digital transmission.

Maish (2003) proposed a precautionary approach to safeguard of children's health from exposure to mobile phone microwave radiation, after the US Environmental Protection Agency (EPA) opinion that children are 10 times more vulnerable than adults to cancer risk from exposure to wide range of chemicals. So if there are adverse health effects from mobile phone use, it will be the children who are in the front line. Banik et al., (2003) published a review explaining the bio-effects of microwaves, concluding that microwave effects were established at all biological levels, from microbial cells to animals as well as the human system and it could a thermally induce different physiological effects.

Zotti-Martelli et al. (2005) reported genotoxic activity, although rather week, of microwave radiation at a frequency falling within the range used in mobile telephone and military systems. Evidences in support of effects of mobile phone microwaves on the *in vitro* exposure of an aqueous solution of electric eel acetylcholinesterase (EeAChE) were given by Barteri et al. (2005) and demonstrated that radio frequency (RF) radiations affect the structural and biochemical characteristics of an important central nervous system enzyme, acetylcholinesterase.

Induction of DNA single-strand and double-strand breaks in human diploid fibroblasts and in rat granulose cell in culture after the exposure of RF-EMF used in mobile phones (1800MHz) was observed by Diem et al. (2005)

The first study showing genetic and most importantly epigenetic alterations on fractionated low-dose radiation exposure in thymus tissue was conducted by Pogribny et al., (2005) and proposed that these changes may play a role in genome destabilization that ultimately lead to cancer.

Incidences of cancer in the areas with elevated levels of natural radiation was studied by Mortazavi et al. (2006) and reported an induction of radioadaptive response in the inhabitants of high background radiation areas and it was claimed that the lack of harmful effects was probably due to the enhanced effectiveness of the repair capacity of the living cells of the inhabitant of these areas. Mothersill and Seymour (2006) reported that adaptive responses, bystander effects and genomic instability, belong to a suite of effects which predominately modulate the low dose response to radiation and these mechanisms were part of the cellular homeostatic response. Adaptive response or radiosensitivity was also studied by Mohammadi et al. (2006) on the blood lymphocytes of inhabitant residing in high background radiation area of Ramsar and induction of adaptive response was reported. Matsumoto et al. (2007) concluded in his study that radiation-induced bystander response may actually contribute to the radioadaptive response.

An increased risk for brain tumour, mainly acoustic neuroma and malignant brain tumour were reported with use of cellular telephones or cordless desktop telephones (Hardell et al., 2006).

Lixia et al. (2006) studied expression of heat shock protein 70 (Hsp70) and cell proliferation of human lens epithelial cells (hLEC) after exposure to the 1.8 GHz radiofrequency field (RF) of a global system for mobile communication (GSM) to investigate the DNA damage and found no difference in the cell proliferation rate between exposed and control cells at any exposure level.

Panagopoulos et al. (2007) investigated the cell death induced by GSM 900-MHz and DCS 1800-MHz mobile telephony radiation in a biological model, the early and mid stages of oogenesis of the insect *Drosophila melanogaster* and suggested the degeneration of large numbers of egg chambers after DNA fragmentation of their constituent cells.

Making use of a more sensitive method of detecting DNA damage involving microsatellite sequences Hardell et al. (2006) observed that exposure of human glioma cells to ELF-EMF gave rise to 3-fold increase in mutation induction compared with unexposed controls. They also suggested that ELF-EMF can also potentiate the mutagenecity of ionizing radiations. On the other hand, Lahkola et al. (2007) denied the increase in risk of glioma in relation to mobile phone use in their study in 5 North European countries.

Hruby et al. (2008) conducted a study to detect whether long-term exposure to 902-MHz GSM-type wireless communication signals would affect 7,12-dimethylbenz(a)anthracene (DMBA)-induced mammary tumours in female Sprague-Dawley rats and observed a significant difference between control and animals exposed to 902-MHz GSM signals.

Ravazzani (2008) proposed that in spite of all the research, on the effects of electromagnetic radiations and their possible health effects, which is going on, there is still much more to achieve. Interpretation of scientific data for use by policy-makers, standard setters, health and environmental authorities and concerned stakeholders must be a continous process, even more, as it is expected that many research projects which are going on at national and international level need to be carefully coordinated to avoid the overlap and help the authorities to achieve the required goals in stipulated time period.

Biological and health effects of electromagnetic field (EMF) have been investigated for many years and work is still going on in this area. With the rapid development of new technologies exposure to both workers and general population to electromagnetic field (EMF) has enormously increased in recent years (Vecchia, 2007). Natural background radiation contributes the vast majority of total population exposure, with radon heading the top of the list. Indoor radon account for the half of all radiation exposure received by the general population (Boice and Lubin, 1997).

It is only during the past decade or so that the techniques of molecular biology have allowed a few glimpses of the basic mechanisms involved in radiation effect. Different interaction mechanisms have been established depending on the nature of the field, and on the frequency. These mechanisms are discussed in detail in various scientific reviews, including WHO's Environmental Criteria Documents (WHO /United Nation Environment Programme/ INRPO, 1984, 1987, 1993), Riley, 1994; ICNRP monographs (Matthews, 2003); Kassam and Andrew (2009); Jiang et al. (2008, 2009); Glaviano et al., (2009).

Ionizing radiations cause damage to cells primarily through rapid generation of burst of reactive oxygen species (ROS) by the direct ionization of macromolecules including DNA (Riley, 1994). Radiation induced ROS are known to participate in damage to various cellular components of the cell, including lipid membrane, proteins and most importantly the DNA, which can lead to carcinogenesis (Petkau, 1987; Halliwell, 1996; Kawanishi et al., 2001) and produce double strand breaks (DSBs) in addition to base damage and single strand breaks (SSBs) (Ames et al., 1993; Glaviano et al., 2009). The main damage to DNA resulting from ROS is the oxidation of the purine and pyrimidine bases and oxidation of guanine residues is the most common giving rise to 7, 8-dihydro-8-oxoguanine (8-OxoG) (Wang et al., 1998). The mutagenic effects of 8-OxoG is very well established, as it can pair efficiently with cytosine or adenine residues (Duarte et al., 1999). Therefore, if an 8-OxoG is established in the genome, it can potentially cause a G: C to a T: A transversion initiating the first step of carcinogenesis (Kovacic and Jacintho, 2001). Single oxidized DNA base changes, produced by electromagnetic radiations, such as 8-OxoG, are removed by base excision repair (BER) and nucleotide excision repair (NER) is responsible for the repair of bulky adducts including radiation induced cyclo-butane pyrimidine dimers (CPDs) and 6-4-photoproduct (6-4PP) (Friedberg, 2001; Kassam and Andrew, 2009).

According to Loscher and Liburdy (1998) even if EMFs do not cause direct DNA damage, there are several other steps in carcinogenesis on which EMFs may act. EMF can effect on signal transduction and intracellular calcium, gene expression and protein synthesis, ornithine decarboxylase (ODC) activity, cell proliferation, and intercellular communication. However, Miyakoshi et al., (1996, 1997) found mutations in the hypoxanthine-guanine phosphoribosyl transferase gene (HGPT) of human melanoma cells exposed to a 400 mT, 50 Hz field. But such extremely strong 50-Hz magnetic fields were not found in occupational or residential environments.

3. GENETIC DAMAGE ASSESSMENT

About 90% of human cancers are carcinomas, perhaps because most of the cell proliferation in the body occurs in epithelia, or because epithelial tissues are most frequently exposed to various form of physical, chemical and radiational damage that favours the development of cancer. Micronucleus assay is a multi-endpoint test of genotoxic responses to clastogens (Kirsch-Volder et al., 2006). The frequency of micronucleus (MN) is extensively used as a biomarker of genomic instability, genotoxic exposure and early biological effect in human biomonitoring studies (Norppa and Falck, 2003). Micronuclei result from chromosome breakage or interference with the mitotic apparatus and such events are thought to be related to carcinogenesis (Bishop, 1987; Cairns, 1975). It is used as an indicator of genotoxic exposition since it is associated with chromosome aberrations (Ramirez and Saldanha, 2002). Buccal cells have been shown to have limited DNA repair capacity relative to peripheral blood lymphocytes, and therefore, may more accurately reflect age-related genomic instability event in epithelial tissue (Dillon et al., 2004). Cytotoxic effects are measured via the proportion of micronucleated cells (MNC), karyolysis (KL), karyorrhexis (KH), broken egg (BE) and bi-nucleated (BN) cells. The comet assay is already recognized as being among the most sensitive methods available for measuring DNA strand breaks (Collins et al., 2003, 2008; Piperakis, 2008; Mohseni-Meybodi et al., 2009). It has further advantageous that observations are made at the level of single cell (Singh et al., 1988, 1994; Tice et al., 1990). The comet tail length, % DNA in tail and Olive moment were mostly used for investigating DNA damage (Tice et al., 1990; Collins, 1992, 2004; Ahuja and Saran, 1999; Armalyte and Zukas, 2002; Collins et al., 2008; Mohseni-Meybodi et al., 2009). The percentage of DNA in the comet tail and tail length are linearly related to DNA break frequency and hence to DNA damage (Collins et al., 2003), so any increase or decrease in the DNA damage will correspondingly lead to increase or decrease in the comet length (μm), comet tail length (μm), % DNA in the comet tail and Olive tail moment in the lymphocyte cells of the population under study. Similarly, comet height (μm), comet area (μm^2), comet intensity, tail area (μm^2), tail intensity and tail moment showed a positive correlation to DNA damage. In case of head diameter, head area, head intensity and % DNA in head showed a negative correlation with DNA damage because lower the damage in DNA, lesser the migration of the DNA from head to tail under the influence of electric field as in the case of controls. On the other hand, comet mean intensity, head mean intensity and tail mean intensity showed a lot of variation in controls as well as exposed population in the present investigation. The comet assay provides a measure of chromatin integrity as a function of time following exposure to a clastogenic agent such as ionizing radiation (Mohseni-Meybodi et al., 2009).

4. MOBILE PHONE USERS

The telecommunication industry is experiencing a robust growth on a global scale. According to recent statistics released by Telecom Regulatory Authority of India on 29th December 2008, the total wireless subscribers (Global System for Mobile Telecommunication i.e. GSM and Code Division Multiple Access i.e. CDMA and others) base stood at 336.08 million and the tele-density reached to 32.34%, at the end of November 2008 (TRAI, 2008). India mainly has two systems of digital mobile telephony i.e. GSM with a carrier frequency around 900 MHz and CDMA with a carrier frequency of 1800 MHz (Repacholi, 2001). Given the immense number of users of mobile phones, even a small adverse effect on health could have a major public health implication (WHO, 2000). Exposure to radio frequency (RF) field in the frequency range of 300 MHz to 2100 MHz emitted by cell phones elicit a host of biological responses and therefore, do represent an unnatural stressor to the biological system no matter how small (Matta and Burkhardt, 2003).

In case of biological system the extent of RF field exposure depend on the amount of energy deposited in tissue, and it is measured by Specific Absorption Rate (SAR) i.e. the amount of energy absorbed per unit time per unit mass of tissue and is expressed in W/kg and its maximum level for modern handsets have been set by governmental regulatory agencies e.g. TRAI (Telephone Regulatory Authority of India), FCC (Federal Communications Commission of USA) and Radiation Protection Bureau, Canada, in many countries (Matta and Burkhardt, 2003). The SAR limit recommended by the International Commission on Non-Ionizing Radiation Protection (ICNIRP), which is 2 W/kg over ten grams of tissue, is highly protective and based on all the available scientific evidences (ICNRP, 1998; WHO, 2005).

5. Individuals Studied

During the present investigations 110 Mobile phone users were investigated. In addition to that, 100 healthy individuals selected at random comprised the controls and matched with the exposed individuals in respect to age, sex and socio-economic status, smoking and alcoholic habits and drug intake, if any but unexposed to any chemical or radiational agent (non-user) and selected at random. Both the blood samples and buccal mucosa cells were obtained from all the 210 subjects. The study was approved by Institutional Human Ethics Committee of Kurukshetra University, Kurukshetra vide letter No. IEC/07/143. Informed consent was obtained from the individual included in the study or from their parents in case of minors. All individuals, exposed and controls, answered a personal questionnaire, from which a profile of each individual was obtained. The subjects also filled in the questionnaire their detailed personal and medical history. Subjects under investigation were all healthy and interviewed to assess their habits. All information was recorded in the proforma designed for the purpose. None of the subject had family history of any genetic anomaly, cancer or major illness nor had they undergone irradiation examination and none have been on drugs for the last 6 months. None of the individuals (exposed and controls) had received chemotherapeutic or cytostatic drugs during the six months prior to buccal mucosa cell and blood collection.

Peripheral blood lymphocytes (PBLs) were chosen for the present study because:

1) PBLs exist in a state of mitotic arrest with lowered capacity for DNA repair and actual damaged DNA can be assessed as it is harder to kill a non-dividing cell.
2) PBLs can accumulate damage which will be expressed when the cells are stimulated to divide *in vitro* or can be studied directly with the help of suitable technique.
3) Bone marrow cells which have been repeatedly exposed to radiations can pass the mutation if any to newly formed PBLs which can be easily studied with the help of SCGE technique, in this way we can also study the accumulating effect of radiations among human populations.

6. Genotoxicity Tests Applied

Single cell gel electrophoresis (SCGE) assay and Micronucleus (MN) assay are used as the biomarkers of choice, for the genotoxic evaluation in the present investigation.

6.1. Micronucleus Assay

During the present study the frequencies of micronucleated cells in the exfoliated buccal epithelial cells of exposed as well as control subjects were investigated. To monitor the cytotoxic effects, karyolysis (KL), karyorrhexis (KH), broken egg (BE) and binucleated cells (BN) were also evaluated in the present study.

The proportion of exfoliated buccal mucosal cells with micronuclei gives the opportunity to assess sensitivity to radiation and genotoxic compounds and in addition to monitor the effectiveness of cancer intervention strategies (Basu et al., 2004; Cerqueira et al., 2008). The study of exfoliated buccal epithelial cells was performed by the standard technique of Tolbert et al. (1992).

The micronucleus assay in exfoliated cells involves examination of epithelial smears to determine the prevalence of cells with micronuclei, extra nuclear bodies composed of chromosome breakage that failed to be incorporated into daughter nuclei at anaphase during mitosis (Schmid, 1975).

6.1.1. Instrumentation and Scoring of Micronuclei

Coded slides were analyzed using a trinocular research microscope at 1000X magnification. At least 1000 cells from each individual were examined and the number of cells with nuclear anomalies was scored. Image acquisitions of representative anomalies as well as normal cells were done with the help of a Digital camera.

Since micronucleus is a rare event, the number of micronuclei missed by using an appropriate method should be reduced as much as possible, and this heavily depends on the type of protocol used for the scoring of micronuclei. The criterion of Tolbert et al. (1992) and Fenech et al. (2003) for scanning cells for micronuclei and other nuclear anomalies were followed during the present study.

Interest in evaluation of cytogenetic damage in epithelial cells has increased recently as the frequency of micronucleus was extensively used as a biomarker of genotoxic exposure and early biological effect in human biomonitoring studies (Norppa and Falck, 2003; Nersesyan and Llin, 2007).

6.2. Single Cell Gel Electrophoresis (SCGE) or Comet Assay

The protocol of Ahuja and Saran (1999) was followed.

6.2.1. Acquisition and Visual Scoring of Images

For comet visualization, Trinocular Research microscope was used at 10X objective. The slides were coded prior to scoring to remove any personal bias. The microscope was pre calibrated using stage micrometer. Manual scoring was done by analyzing one comet at a time at initial stages then followed by the use of freely available Comet imaging softwares i.e. CometScore™ which is a freeware version of the AutoComet™ - Automatic Comet Assay System (http://autocomet.com).

From each replicate 100 cells were randomly chosen (50 from each duplicate slide), and analyzed under microscope and image acquisition was done. The saved image files were later on retrieved by imaging software for analysis. The relationship between light intensity that a camera detects and the gray value of the pixel on the image is linear. Hence, the relationship between the amount of DNA present in one cell in the comet and the corresponding pixel gray value is linear (Kumaravel et al., 2007). Geometric calibration of the image analysis system was also performed so that software was able to measure the comets in realworld units (Vilhar, 2004). The scoring was done with the help of CometScore™, software made available by the TriTek Corporation, an automatic image acquisition and cross-platform data analysis system designed specifically for the high-throughput comet assay. The scoring was done by single observer to minimize the inter observer variability. Comet can be divided into head and tail regions.

6.2.2. Comet Parameters

For Damage Index (DI) calculation, following comet parameters were included in the study: Comet length, Comet height, Comet area, Comet intensity, Comet mean intensity, Head diameter, Head intensity, Head mean intensity, % DNA in head, Tail length, Tail area, Tail intensity, Tail mean intensity, % DNA in tail, Tail moment and Olive tail moment.

Out of these parameters Comet length, Tail length, Tail moment and Olive moment were more reliable and mostly used by various investigators for studying DNA damage (Tice et al., 1991; Collins, 1992, 2004; Ahuja and Saran, 1999; Armalyte and Zukas 2002; Kumaravel and Jha, 2006). Comet length is defined as the total distance between the two ends of the comet or can also be defined as the number of pixels in horizontal direction in comet. Comet height is the vertical distance of the comet or can also be defined as the number of pixels in vertical direction in comet. Comet area is total number of pixel in comet. Comet intensity is sum of the pixel intensity values in comet. Comet mean intensity is mean intensity of pixels in comet. Head diameter can be defined as the number of pixel in horizontal direction in head. Head area is equal to number of pixel in head of the comet. Head intensity refers to the sum of pixel intensity values in head. Head mean intensity equals to the mean intensity of pixels in head and % DNA in head can be defined as the total head intensity divided by total comet intensity (multiplied by 100). Tail length is calculated by subtracting the head diameter from comet length. Tail area is the number of pixels in tail. Tail intensity is the sum of the pixel intensity values in tail. Tail mean intensity refers to mean intensity of pixel in tail and % DNA in tail is defined as the total tail intensity divided by total comet intensity (multiplied by 100). Tail moment is calculated by multiplying %DNA in tail by tail length and Olive tail moment is summation of tail intensity profile values multiplied by their relative distance to the head center, divided by total comet intensity. Migration length is directly related to fragment size and is expected to be proportional to the extent of DNA damage (Kumaravel and Jha, 2006). The process of repair in irradiated nuclei as well as in control samples can be stopped by keeping the nuclei suspension at a temperature of 4°C (Armalyte and Zukas, 2002).

6.3. Statistical Analysis

The data were subjected to Student t-test, correlation co-efficient and ANOVA. These statistical methods were applied using the computer software SPSS 11.5 and ORIGIN 6 as well as Microsoft Office Excel on the data obtained during the present study and graphical representations of the observations and results.

7. PROFILE OF THE SUBJECTS STUDIED

The general characteristics of the exposed as well as control subjects were studied/compared w.r.t. age, sex, smoking, drinking and dietary habits etc. (Table 1). Health problems such as eye ailments; nausea, hypertension and cancer were reported negative in the concerned population under investigation. However, regarding the health effects 4 subjects were complained to have head ache and giddiness.

About 23% of the total subjects were residing near the 500 meter of the proximity to the base station i.e. mobile phone transmission towers (Table 1).

Table 1. General characteristics of study group and controls

Sr. No.	Variables		Total	Exposed	Control
1	n		210	110	100
2	Average age (years)		23.82	23.95	22.38
3	Age range		18-36	18-32	18-36
4	Sex	Males	92 (43.8%)	49 (44.5%)	43 (43%)
		Females	118 (56.2%)	61 (55.5%)	57 (57%)
5	Smoking habits	Non-smokers	201 (95.7 %)	102(92.7%)	99 (99%)
		Smokers	9 (4.3 %)	8 (7.3%)	1 (1%)
6	Drinking habits	Non-alcoholics	195 (92.9 %)	96 (87.3%)	99 (99%)
		alcoholics	15 (7.1 %)	14 (12.7%)	1 (1%)
7	Dietary habits	Vegetarian	171 (81.4 %)	78 (70.9%)	93 (93%)
		Non-vegetarian	39 (18.6 %)	32 (29.1%)	7 (7%)
8	Average duration of exposure (Years)		-	2.53	0
9	Average daily exposure (minutes)		-	70.78	0
10	Subjects proximity to Base station (500m)		49 (23.3%)	47 (42.7%)	2 (2 %)
11	Education (n=210)		UG (4.3%), G (43.8%), PG (51.9%)		
12	Population profile (n=210)		Student (n=199), teachers (n=4) business men (n=2), housewives (n=5)		
13	Health effects/ drug intake		Skin disease (n=1), Respiratory problems (n=3), head ache (n=4), Antibiotic taken (n=3)		

All details regarding the use of mobile phone, duration of exposure in term of average duration of exposure (years) as well as duration per day were recorded (Table 1). Average mean daily frequency of numbers of calls was 15.55 (range from 1 to 80) with an incoming of 8.46 and outgoing of 7.03 average calls. The average mean daily duration of calls comes out to be 70.78 minutes (range from 5 to 540 minutes) with an incoming of 42.14 minutes and outgoing of 29.37 minutes. The mobile phone was kept at 'on mode' for 24 hours by 105 (95.5%) of the users, rest of the users (4.5%) kept it on for various duration ranging from 12 to 18 hours per day.

7.1 Nuclear Anomalies

The following points were kept in consideration while proceeding for the micronuclei assay:

a) There is period of approximately 10-14 days between micronucleus formation in the basal level and migration of the cells to the surface, where they can be exfoliated and then studied (Rosin 1992). Hence, it was assumed that the effect shown by the cells was actually due to the exposure occurred in the past 10-15 days.
b) Washing of the epithelial cell with 1N HCl to eliminate the microbes which can resemble micronuclei in epithelial cells.
c) Staining of the epithelial cells with DNA specific stains (e.g. Feulgen, acridine orange or aceto-orcein) in order to recognize actual micronuclei, but not artifacts (Casartelli et al., 1997).
d) The scoring was done by single observer to minimize the inter observer variability because, about half of the variability in the MNC counts within exposure groups was due to intra- and inter-observer variability rather than intrinsic variability in true MNC frequency between scrapings (Tolbert et al., 1992).

In viewing sample slides, it became readily apparent that there were other phenomena occurring in the cells, some of which were difficult to distinguish from classical micronuclei. Normal buccal epithelial cell showed clear complete nuclei (Fig. 1). Cell debris can interfere with the scoring of nuclear anomalies so care had been taken to avoid that by following the well established criteria (Tolbert et al., 1992). During the present study following nuclear anomalies were observed in sample smears in addition to micronuclei (Fig. 2).

1) Broken egg (BE), in rare cases, some cells may be blocked in a binucleated stage or may exhibit nuclear buds also known as "Broken egg" (Fig. 3), a biomarker of gene amplification, or nuclei that appear cinched with a differential stain.
2) Binucleates (BN), or the presence of two nuclei within a cell (Fig. 4).
3) Karyorrhexis (KH), or nuclear disintegration involving loss of integrity of the nucleus (Fig. 5).
4) Karyolysis (KL), or nuclear dissolution, in which a stain-negative, ghost-like image of the nucleus remains (Fig.6).

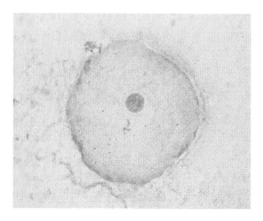

Figure 1. Photograph of exfoliated epithelial cell of buccal mucosa showing a normal nucleus (1000X).

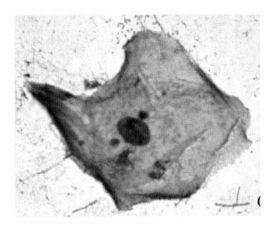

Figure 2. Photograph of exfoliated epithelial cell of buccal mucosa showing two micronuclei (1000X).

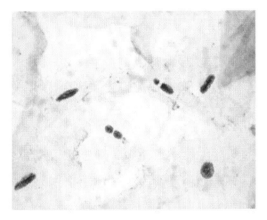

Figure 3. Photograph of exfoliated epithelial cells of buccal mucosa showing two broken eggs (1000X).

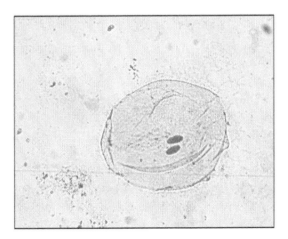

Figure 4. Photograph of exfoliated epithelial cell of buccal mucosa showing binucleate condition (1000X).

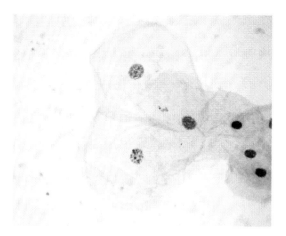

Figure 5. Photograph of exfoliated epithelial cells of buccal mucosa showing karyorrhexis (1000X).

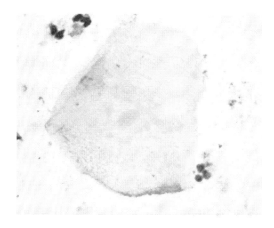

Figure 6. Photograph of exfoliated epithelial cell of buccal mucosa showing karyolysis (1000X).

In addition to these there were other phenomenon occurring in the cells such as *pycnosis* or shrunken nuclei, *condensed chromatin* in which nuclear chromatin appear aggregated, degenerated nuclei, three lobed nucleus, double micronuclei were also observed during scoring of the anomalies. However, these were not counted in the anomalies. Pycnosis and condensed chromatin are considered as a part of normal epithelial cell differentiation and maturation.

The results of the cytological observations on MNC, TMN, KL, KH and BE in terms of mean, standard errors (S.E.), standard deviations (S.D.) of means and range both in exposed and control subjects are shown in Table 2. The mean frequency of MNC (9.227±0.608) and TMN (10.127±0.715) in exposed group showed a significant difference at 0.01 levels with respect to controls where MNC and TMN were 3.500±0.290 and 3.760±0.305, respectively (Fig. 9).

The mean of KL in controls was 8.720±0.880 and in exposed subjects was 12.272±1.418. The value of means of KH in exposed subjects (1.909±0.398) slightly lower than in controls (2.980±0.689). Mean broken egg was found to be more in controls (0.960±0.309) as compared to exposed subjects (0.518±0.215). Frequency of presence of more than one nucleus in a cell (binucleated) was significantly higher in exposed (2.245±0.307) in comparison to controls (0.440±0.087). The difference was not found statistically significant for KL, KH and BE when comparison was done between exposed and control subjects. Cells with MN count more than one were predominant in the exposed population apparently as a result of the clastogenic effect of electromagnetic radiations emitted from mobile phone antennas, on the other hand in control cells two or more MN per cell were lower than in exposed group.

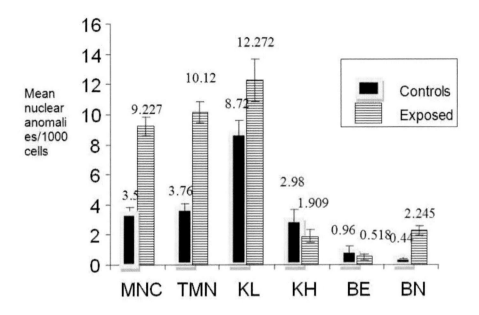

Figure 9. Cytological observations of exposed (n=110) and control (n=100) subjects belonging to the study group.

Cytological observations were also made with respect to smoking and drinking habits and sex of the subjects (Table 2).

Cytological observations in relation to duration of exposure (years) to mobile phone radiation of exposed subjects are shown in Fig. 10. It is of interest that a positive correlation was found between duration of mobile phone use and MNC and TMN frequency, showed gradual increase, although the correlation was not found to be statistically significant. During the initial years (1–3) of mobile phone use, MNC and TMN frequency showed a gradual increase. Pearson correlation between 0–1, 1–2, 2–3 and 3–4 years of exposure has been calculated and found to be positively correlated for TMN. Surprisingly, more than 3 and 4 year's exposure showed a slight decrease in MNC and TMN frequency (Fig. 10).

Pearson correlation between various nuclear anomalies observed in exposed population was found to be statistically significant at 0.01 level in case of MNC, TMN and KH; statistically significant correlation at 0.01 level, was also found between TMN vs. KL, KH and MN>1; KL vs. KH and BN; KH vs. MN>1; BE vs. BN.

Table 2. Cytological observations of subjects belonging to the study group

Sr. No.	Group	Subjects	MNC	TMN	KL	KH	BE	BN
1	Non-smoker	Controls	2.65±0.565**	2.82±0.598**	10.55±2.091	2.31±0.972	0.93±0.454	0.71±0.256**
		Exposed	9.08±0.638**	9.97±0.757**	12.25±1.448	1.92±0.416	0.56±0.232	2.35±0.326**
2	Smoker	Controls	10.00±0.000	10.00±0.000	22.000±0.000	2.000±0.000	0.00±0.00	4.00±0.00
		Exposed	12.88±2.048	14.75±1.968	14.75±6.958	2.50±1.547	0.13±0.125	1.50±8.24
3	Non-alcoholic	Controls	3.434±0.285**	3.697±0.301**	10.55±2.091	2.31±0.972	0.93±0.454	0.71±0.256**
		Exposed	9.15±0.663**	10.14±0.780**	12.40±1.517	1.94±0.436	0.59±0.246	2.51±0.341**
4	Alcoholic	Controls	10.00±0.000	10.00±0.000	22.000±0.000	2.000±0.000	0.00±0.00	4.00±0.00
		Exposed	10.785±1.644	11.214±1.950	12.71±4.310	2.140±1.016	0.07±0.071	0.86±0.501
5	Male	Controls	3.24±0.432*	3.6±0.461*	8.220±1.383	6.040±1.390*	2.000±0.654*	0.09±0.089*
		Exposed	9.13±0.729*	10.05±0.850*	11.570±1.563	1.930±0.467*	0.28±0.100*	2.18±0.433*
6	Female	Controls	3.71±0.393*	3.89±0.410*	9.67±1.132	0.47±0.187*	0.11±0.062	0.80±0.143*
		Exposed	9.35±1.029*	10.22±1.218*	13.14±2.53	1.88±0.686*	0.82±0.466	2.33±0.434*

** Significant at 0.01 level (Student's "t" test)

* Significant at 0.05 level (Student's "t" test)

MNC-micronucleated cell, TMN-total micronuclei, KL- karyolysis, KH-karyorrhexis, BE-broken egg, BN- binucleated cells.

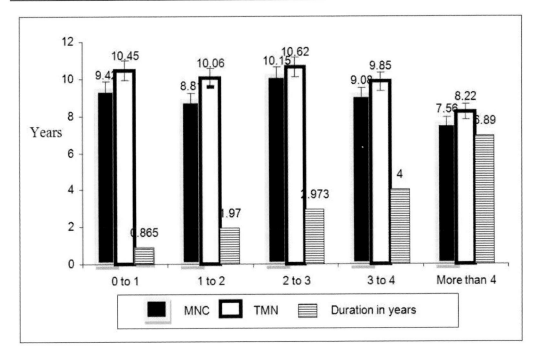

Figure 10. Cytological observations in relation to duration of exposure (years) to mobile phone radiation of exposed subjects of the study group.

Statistically significant correlation at 0.01 level, was found between duration of call (minutes) vs. MNC, TMN, KH, MN>1 and positive correlation was found in case of KL, BE and BN. Statistically significant correlation at 0.01 level was found between duration of use of mobile phone (years) and BE. Positive correlation was observed between duration of use of mobile phone (years) and KH and BN, however, negative correlation was found in relation to MNC, TMN and KL.

Comparison was also made with respect to the nuclear anomalies observed and presence or absence of telecommunication base (mobile base antenna) in 500 meters of the living habitat of the subjects. Significant difference was observed between the nuclear anomalies in subject belonging to two sub-groups i.e. (A) living near base station and (B) not living near base station (Table 3).

Table 3. Assessment of genetic damage in relation to proximity to mobile base station and nuclear anomalies in subjects of study group

Sr. No.	Nuclear anomalies	Mobile phone base station [more than 500 m]	Mobile phone base station [with in 500 m]
1	MNC	5.61±0.422*	9.43±0.886*
2	TMN	6.01±0.484*	10.65±0.999*
3	KL	10.27±0.995	12.20±1.686
4	KH	2.57±0.475	1.92±0.600
5	BE	0.83±0.238*	2.47±0.458*
6	BN	1.08±0.178	0.39±0.139
7	MN>1	0.24±0.076*	0.78±0.223*

*Significant at the 0.05 level (Student's "t" test)

7.2. DNA Damage

The influence of the RF-EMF (Radio frequency- Electromagnetic field) exposure on the generation of DNA strand breaks in lymphocyte cells derived from the human blood samples was evaluated using the alkaline comet assay.

The results of the comet assay for Damage Index calculation, in the present investigation were recorded in terms of following parameters i.e. Comet length, Comet height, Comet area, Comet intensity, Comet mean intensity, Head diameter, Head intensity, Head mean intensity, % DNA in head, Tail length, Tail area, Tail intensity, Tail mean intensity, % DNA in tail, Tail moment and Olive tail moment in terms of mean, standard error (S.E.), standard deviation (S.D.) of means and range both in exposed and control subjects. Migration length is generally believed to be related to fragment size and would be expected to be proportional to the level of single strand breaks and therefore, DNA damage on the same ground inversely proportional to the extent of DNA cross-linking (Tice et al., 2000).

In addition to other parameter mentioned above comet types were also used to study the genotoxic effects by various workers in which cells were sorted out into four classes based on the tail size and shape i.e. 0-4 types (Collins, 2004) (Fig. 7-8).

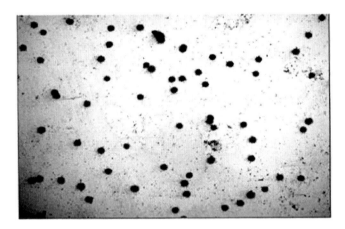

Figure 7. Lymphocyte cells from control subject showing normal cells.

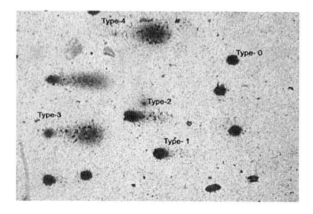

Figure 8. Lymphocyte cells from exposed subjects showing five types of comets.

Genetic Damage in Exposed Subjects

Almost all comet parameters viz. the mean frequency of comet length (μm) (90.860±0.469), comet height (μm) (44.626±0.110), comet area (μm^2) (3129.030±24.029), comet intensity (271143.56±2066.974), comet mean intensity (126.187±0.356), head diameter (μm) (43.133±0.097), head area (μm^2) (1768.747±10.930), head intensity (161331.03±986.588), head mean intensity (129.88±0.356), % DNA in head (69.455±0.207), tail length (μm) (47.728±0.441), tail area (μm^2) (1360.283±18.958), tail intensity (109812.52±1579.021), tail mean intensity (315.016±43.140), % DNA in tail (30.544±0.207), tail moment (19.305±0.284) and Olive tail moment (13.943±0.194) in exposed subjects of Group - I (Mobile phone users) population under study showed a significant difference at 0.05 levels in comparison to controls (Fig. 11).

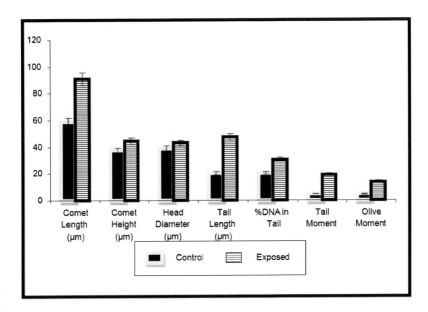

Figure 11. Comet assay observations of control (n=100) and exposed (n=110) subjects belonging to the study group.

Genetic Damage in Smoker Exposed and Non-Smoker Exposed Subjects

In smoker exposed subjects certain comet parameters viz. comet mean intensity (114.395±11.148), head diameter (46.366±2.025), head area (1813.901±128.153) and % DNA in head (71.0±4.09) showed higher values in comparison to that in non-smoker exposed subjects.

Genetic Damage in Alcoholic Exposed and Non-Alcoholic Exposed Subjects

Most of the comet parameters in non-alcoholic exposed group depicted higher values in comparison to alcoholic exposed group except for head diameter and % DNA in head.

Genetic Damage in Relation to Gender of the Subjects

To find out any kind of relationship between the sexes of the subjects and DNA damage among the study group, a thorough comparison was done. Female subjects (n=118) were found to have more comet length, comet area, comet intensity, comet mean intensity, head mean intensity, tail area, tail intensity, % DNA in tail, tail length, tail moment and Olive moment as compared to male subjects in exposed groups. Similar correlation was also observed when comparison was made between the control female and control male subjects (Table 4).

Table 4. Assessment of genetic damage in exposed (male = 49, female = 61) and control (male = 43, female = 57) subjects of the study group

Sr. No.	Comet parameter		Male Mean±SE	Female Mean±SE
1	Comet Length (μm)	Exposed	89.831±5.783	94.014±6.287
		Control	55.468±2.319*	61.775±2.000*
2	Comet Height (μm)	Exposed	44.630±1.106	44.870±1.535
		Control	37.351±0.417	37.676±0.575
3	Comet Area (μm²)	Exposed	3056.397±280.276	3285.980±337.414
		Control	1612.33±88.876	1872.332±93.969
4	Head Diameter (μm)	Exposed	43.595±0.782	42.297±0.721
		Control	39.159±0.574	39.049±0.644
5	Head Area (μm²)	Exposed	1769.461±92.270	1716.981±95.289
		Control	1244.588±43.561	1405.899±63.486
6	%DNA in Head	Exposed	70.605±2.355	67.205±2.103
		Control	82.088±1.344*	78.349±1.230*
7	Tail Length (μm)	Exposed	46.240±5.423	51.716±5.829
		Control	16.308±1.890*	22.726±1.673*
8	%DNA in Tail	Exposed	29.394±2.355	32.794±2.103
		Control	17.911±1.344*	21.650±1.230*
9	Tail Moment	Exposed	18.638±3.456	22.225±3.740
		Control	4.149±0.724	5.370±0.646
10	Olive Moment	Exposed	13.439±2.346	16.0625±2.594
		Control	4.043±0.546	5.143±0.497

*Significant at 0.05 level (Student's "t" test).

Genetic Damage in Relation to Dietary Habits

In exposed vegetarian subjects, the mean values of different parameters under study were found higher than in non-vegetarian exposed subjects except for head diameter, head area and % DNA in head. While vegetarian control subject were found to have lower degree of basal DNA damage as compared to non-vegetarian control subjects.

In the present investigation, in mobile phone users (exposed) the mean frequency of MNC, TMN and BN showed an increase and the difference was found to be statistically significant at 0.01% levels with respect to controls, similarly the frequency of BN was also significantly higher in case of exposed population as compared to controls indicating genotoxic effects of mobile phone radiation in the exposed population.

Similar results were found in other reports which showed an increase in MN frequency; following *in vitro* exposure of human lymphocytes to 1800 MHz microwave radiation at a power density of 5, 10 and 20 MW/cm^2 (Zotti-Martelli et al., 2005). Our results are in substantial agreement with the previous reports on the induction of micronuclei and chromosome damage by non-ionization radiation. A significant increase in MN frequency was reported by Carborari et al. (2005).

However, some studies showed no effects on DNA damage, chromosomal aberrations or micronuclei in human peripheral blood lymphocytes or in rat peripheral blood or bone marrow cells exposed to two mobile phone signals (CDMA at 847.74 MHz, FDMA at 835.62 MHz) or to 2.45 GHz microwaves (Vijayalaxmi et al., 2000, 2001).

During the present investigation for the comet assay, statistically significant difference was observed in the entire comet parameters between exposed and control population thereby indicating DNA damage in the form of comet length, tail length, % DNA in tail and Olive tail moment showed significant increase in exposed subjects. In the present study, direct relationship was found between the time of exposure and DNA damage. In case of controls where there was no exposure, only minor basal damage was recorded which was indicated by the smaller comet tail length of 20.095 ± 0.164 μm as compared to 47.728 ± 0.441 μm in exposed subjects.

Similarly, Ramirez and Saldanha (2002) reported that exposure of cultured cells to mobile phone RF energy caused DNA damage. Lai and Singh (1997) reported that acute exposure of rats to 60 Hz magnetic fields up to 0.5 mT increases DNA strand breaks in brain cells. Bohr et al. (1997) suggested that the exposure to high frequency EMFs may alter the conformation of the molecules and their ability to function and a breaking of the chain can result.

The latest Interphone study, which was one of the largest epidemiological study of mobile phone radiations and tumours (Cardis et al., 2007) on the risk of glioma (brain cancer) in five northern European countries from cellphone use for a period of 10 years or more yielded a significant increased risk of glioma on the side of the head where the mobile phone was used (Lahkola et al., 2007). Earlier, Schoemaker et al. (2006) also reported increased risk of acoustic neurinoma related to durations of use of 10 years or more.

Recently Yadav and Sharma (2008) reported an increase in the frequency of micronuclei in the subjects exposed to mobile phone radiations as compared to control subjects.

Maes et al. (1996) found a synergistic effect in the case involving mitomycin C in their investigation of 954-MHz waves emitted by the antenna of a GSM base station (15 W power output, SAR= 1.5 w/kg). However, when cells were exposed to waves of 935.2 MHz (4.5W) followed by mitomycin C, synergism was much less evident.

Russian National Committee on Non-Ionizing Radiation (2008) gave the opinion that following health hazards are likely to be faced by the children mobile phone users in the nearest future: disruption of memory, decline of attention, sleep disorders, increase in the sensitivity to the stress and increased epileptic readiness. The remote health risks include the brain tumours, Alzheimer disease and other type of nervous degeneration in the brain.

However, some of the findings reported that use of mobile phone was not associated with an increased risk of acoustic neuromas (Muscat and Malkin, 2002) and there was no statistically significant association of mobile phone use with overall incidence of brain cancer or the incidence of salivary gland cancer (Auvinen and Hietanen, 2002). Regular use, or long-term (greater than 10 year) use, or heavy (greater than 500 hours) use of mobile phones (analog or digital) was not associated with an increase in the incidences of malignant brain tumor (Lonn and Ahloom, 2005).

Lantow et al. (2006) reported no effects either on release of reactive oxygen species or expression levels of heat shock protein HSP70 in primary human monocytes and lymphocytes due to radio frequency exposure.

In 2005 the WHO published a fact sheet concluding that "there is no convincing scientific evidence that the weak RF (radio frequency) signals form wireless network cause adverse health effects".

Huuskonen et al. (1998) also reviewed both epidemiological and experimental evidence on the possible role of ELF (Extremely low frequency) and VLF (very low frequency) magnetic fields in teratogenic and reproductive outcome and indicated that exposure to low frequency magnetic fields during pregnancy does not exert strong effects on embryonic development.

Lai and Singh (1996), using comet assay or single cell gel electrophoresis, have reported DNA strands break in rat brain cells irradiated *in vivo* with 2450 MHz radiation at SAR of 0.6 and 1.2 W/kg. Radio frequency exposure of cells *in vitro* has been linked to changes in transcription and cell proliferation assayed by the incorporation of an RNA-precursor and a DNA-precursor, respectively. RF exposure has also been linked to change in cell cycle (Zeni et al., 2005).

In previously published studies with cultured human lymphocytes no induction of chromosome aberrations, micronuclei or sister chromatid exchange was measured after exposure to a 900 MHz GSM signal (SAR between 0.2 and 10 W/kg) (Zeni et al., 2005, Scarfi et al., 2006). Evidence from *in vitro* studies to date is consistent with those epidemiological results which state that there is no statistically significant correlation between cancer and exposure to cell phone field (Blockwell et al., 2002). Similar experiments on rats exposed to one of the mobile signals (FDMA or CDMA) were performed but no statistically significant difference was found between the exposed animals for any tumour in any organ (Regina et al., 2003). Other workers also reported the lack of genotoxic effect of radio frequency level as evident from experimental study on animals at the cellular level (Krewski et al., 2001; Mulder et al., 2002).

William et al., (2008) investigated the effects of mobile-phone like RF exposure on subjects with symptom related to mobile phone and found no effects of RF exposure in their study.

The mean frequencies of other nuclear anomalies did not show any significant difference; however, the outcome frequencies were contemporary to that reported by other authors (Stich et al., 1982, 1984; Tolbert et al., 1992; Fenech and Crott., 2002; Thomas et al., 2008). Regional variations in the mean percentage of micronucleated cells in the buccal mucosa of controls in Orissa population (0.39%) and that of Meghalaya (0.44%) have been reported (Stich et al., 1982, 1984).

Taking into consideration the comparative aspects of smoking and drinking habits as well as gender of the subject, significant difference was observed in MNC, TMN and BN in case of control and exposed population (Table 2). Females showed higher values of MNC, TMN and KL as compared to males of respective group; this was in agreement with Piyathilake et al. (1995) who reported 2.8 fold higher frequencies of micronucleated cells in females as compared to males. On the other hand, our finding was in contrast to reports of Benites et al. (2006) who studied Brazilian population and found significantly higher number of micronuclei in males. Similar finding was reported by Nishioka et al. (1981). The effects of alcohol consumption, age and gender in micronuclei formation has been studied in several earlier investigations, and the results were highly controversial (Sarto et al., 1987; Stich and Rosin, 1983; Benites et al., 2006; Speit and Schmid, 2006). Stich and Rosin (1983) reported a pronounced combined effect of alcohol in combination with smoking, whereas in another study an inverse relation was found (Ramirez and Saldanha, 2002).

During the present investigation similar comparative aspects in case of smoking and drinking habits as well as gender of the subjects in terms of comet analysis also showed significant difference between non-smoker exposed and non-smoker control and non-alcoholic exposed vs. non-alcoholic control. The comparisons between smoker exposed and smoker control, smoker exposed vs. non-smoker exposed, alcoholic exposed vs. non-alcoholic exposed and alcoholic exposed vs. alcoholic control group did not show significant difference. Some of the parameters in non-smoker exposed were found to have higher values than in smoker exposed subjects. This may perhaps be due to the possibility of the former being exposed to passive smoking. Also the sample size of smoker exposed subject (n=8) was quite less as compared to non-smoker exposed subject (n=102). Ramirez and Saldanha (2002) also reported a combined effect of alcohol in combination with smoking and opined that it may lead to an increase in the DNA damage in alcoholic and smoker subjects. The comparison in terms of sex and dietary habits also revealed interesting information. In female subjects more DNA damage was observed both in control and exposed group as compared to males (Table 4) which was in contradiction to earlier reports (Nishioka et al., 1981; Piyathilake et al., 1995). Significant differences were also observed for all the parameters except for comet head and tail mean intensity in exposed female vs. control females and exposed male vs. control male, which strengthen the view that mobile phone radiation cause the DNA damage in the exposed group in both sexes. Some studies have postulated a connection between prolonged use of cellular phones with both brain cancer and leukemia (Hardell et al., 2002; Grove et al., 2002). On the other hand, an *in vitro* study reported no effect on the frequency of neoplastic transformation after exposure to cell phone RF-field (835 MHz and 848 MHz) at 0.6 W/kg for 7 day in cell culture (Roti et al., 2001). Diem et al. (2005) reported in their study that mobile-phone modulations RF-EMF induces single strand breaks in human diploid fibroblasts and in rat granulose cells in culture.

Interpretation of the results in relation to dietary habits of the subjects also confirmed the genotoxic effects of mobile phone radiations on comparison between same dietary habits group i.e. vegetarian exposed vs. vegetarian control; the former showed higher DNA damage as revealed by different comet parameters. The difference was found to be significant. The non-vegetarian control showed a higher value of DNA damage as compared to vegetarian control. DNA damage was also observed higher in vegetarian exposed as compared to non-vegetarian exposed and non-vegetarian exposed as compared to non-vegetarian control, but the difference was not found to be significant. Cross-correlation analysis between the biomarkers of micronucleus assay showed one of the interesting observations during the present study that there was a positive correlation between various nuclear anomalies. Thus the possibility of a similar mechanism of their formation is more likely proposition. As till now mechanism of formation of broken egg is not known but having a positive correlation with BN, so there may be some kind of similarity in the mechanism of their formation. A similar kind of correlation between KH and condensed chromatin was also reported by Thomas et al. (2008).

Positive correlation was observed between the duration of exposure in terms of duration of calls (minutes) and frequency of MNC, TMN and KH at o.o1 level and with MN>1 at 0.05

level. These findings were in accordance with the earlier findings (Schoemaker et al., 2006). Strangely enough, a negative correlation was observed in nuclear anomalies and duration of use in terms of years; although during initial years of exposure a positive correlation was observed but after longer exposure negative correlation was obtained. This was in accordance with Yadav and Seth (2000) which reported a decrease in DNA damage and micronuclei formation with increase in duration of exposure in years. It may be possible that some kind of adaptive mechanism was developed by the body to counter act the radiation insult.

An important finding in the present study is that the duration of exposure and frequency of MN and TMN frequency showed a positive correlation in initial years of exposure i.e. 0-1, 1-2 and 2-3 years but then a slight decrease in the frequency of MNC and TMN was observed for subjects exposed up to 3-4 years and more than 4 years. It may be possible that cellular DNA repair mechanism may have got activated more in subjects who were exposed to prolonged use of low frequency EMR *in vivo* conditions, which worked to protect DNA. But it is highly conjectured at this stage and exact reason is not known.

It is conjectured that some clandestine mechanism helps DNA to develop some sort of adaptability to protect itself, however, in case of continuous exposure to week genotoxic agents such as low frequency EMR, it is a matter of further investigation.

Although scientific evidence for health hazards of low level cell phone radiation is weak, the World Health Organization has recommended that the precautionary principle could be voluntarily adopted in this case (http://www.who.int/docstore/peh-emf/publications/facts_press/EMF-precaution.htm).

CONCLUSION

In the present study, significant degree of genetic damage in the population exposed to electromagnetic radiations was observed. Population exposed to mobile phone radiations as well as X-ray radiations showed a higher degree of genetic damage in DNA as compared to control group. Positive correlation was also observed between the duration of exposure and genetic damage. However, on long duration of exposure, a negative correlation was observed. It may be possible that body develops some kind of adaptability to deal with the genetic damage caused by the electromagnetic radiations. Gender wise difference in relation to effects of radiation on DNA was also observed. Higher degree of genetic damage was noticed in female subjects as compared to males. A correlation was also observed between broken egg (BE) and Bi-nucleation (BN), it is proposed that there may be some sort of similarity in the mechanism of their formation.

However, concerns about long-term effects of chronic exposure to low-level EMF have created a demand for precautionary measures beyond the standard for recognized, acute effects. Development of safety guidelines is a dynamic process that evolves with the progress of knowledge. ICNRP continuously checks the validity of its recommendations by monitoring both the advancement in research on biological and health effects of electromagnetic fields, and the development of emerging technologies that may involve the introduction of new modalities of exposure. While there seems to be some need to think about the basic restrictions and reference levels, an update of the scientific rationale that include the most recent research findings is need of the hour. The urgency of the need is not in dispute.

However it is important that caution must be exercised in making any general conclusion based cytological and molecular studies on the clastogenicity, genotoxicity and heritable effects of electromagnetic radiations.

FUTURE PERSPECTIVES

Since large section of modern society is exposed to various types of occupational/ work/ life style hazards the primary objectives of the future studies should be to:

1) provide analytical, quantitative, and theoretical information on the induction and processing of different patterns of complex DNA damage (DSB and non-DSB clustered DNA lesions) produced directly and indirectly (bystander effects by EMF radiations *in vivo* and *in vitro;*
2) measure the biological effectiveness of high and low frequency radiations for the induction of chromosome/ DNA damage and apoptosis *in vivo* and *in vitro* and correlate these end points with the levels of complex DNA damage detected in each case;
3) improve the experimental detection methods for including the very small DNA fragments expected to be induced in the case of EMF radiations. A better knowledge of the induction and repair of complex DNA damage is essential for the calculations of the risk factors of electromagnetic radiations to induce carcinogens.

Anti-oxidant treatments in animals and humans could be beneficial in preventing or reducing some complications of microwave radiations. Although present scientific information does not indicate the need for any special precautions for use of mobile phones, as far as individuals are concerned, they might choose to limit their own EMF exposure by limiting the length of calls, or using "hands-free" devices to keep mobile phone away from head and body.

REFERENCES

Agarwal A, Deepinder F, Sharma RK, Ranga G and Jianbo Li (2008) *Effect of cell phone usage on semen analysis in men attending infertility clinic: an observational study.* Fertility and Sterility 89:1, 124-128.

Ahlbom A, Green A, Kheifets L, Savitz D and Swerdlow A (2004) *Epidemiology of health effects of radiofrequency exposure.* Environmental Health Perspectives. 112: 1741-1754.

Ahmed RG (2005) *Damage pattern as a function of various types of radiations.* Medical Journal of Islamic World Academy of Sciences 15: 4, 135-147.

Ahuja YR and Saran R (1999) *Alkaline single cell gel electrophoresis assay I. Protocol.* J. Cytol. Genet. 34:1, 57-62.

Aitken RJ, Bennetts LE, Sawyer D, Wiklendt AM and King BV (2005) *Impact of radiofrequency electromagnetic radiation on DNA integrity in the male germ line.* Int. J. Androl. 28: 3, 171-179.

Ames BN, Shigenaga MK and Hagen TM (1993) *Oxidents, antioxydents and degenerative disease of aging.* Proc. Natl. Acad. Sci. USA 90: 7915-7922.

Basu A, Ghosh P, Das JK, Banerjee A, Ray K and Giri AK (2004) *Micronuclei as Biomarkers of Carcinogen Exposure in Populations Exposed to Arsenic Through Drinking Water in West Bengal, India: A Comparative Study in Three Cell Types.* Cancer Epidemiology Biomarkers and Prevention 13: 820-827.

Antoccia A, Degrassi F, Battistoni A, Ciliutti P and Tanzarella C (1991) *In vitro micronucleus test with kinetochore staining: evolution of test performance.* Mutagenesis 6: 319-324.

Armalyte J and Zukas K (2002) *Evaluation of UVC-induced DNA damage by SCGE assay and its repair in barley.* Biologija. 2: 26-30.

Auvinen A and Hietanen M (2002) *Brain tumors and salivary glands cancers among cellular telephone user.* Epidemiology 13: 356-359.

Balzano Q, Garay O and Manning T (1995) *Electromagnetic energy exposure to stimulated users of portable cellular telephones.* IEEE Trans. Vehic. Tech. 44: 390-403.

Banik S, Bandyopadhyay S and Ganguly S (2003) *Bioeffects of microwave- a brief review.* Bioresource Technology 87: 155-159.

Barteri M, Pala A and Rotella S (2005) *Structural and kinetic effects of mobile phone microwaves on acetyllcholinesterase activity.* Biophysical Chemistry 113: 245-253.

Belien JAM, Copper MP, Braakhuis BJM, Snow GB and Baak JPA (1995) *Standardization of counting micronuclei: definition of a protocol to measure genotoxic damage in human exfoliated cells.* Carcinogenesis 16: 2395-2400.

Benites CI, Amado LL, Vianna RA and Martino-Roth Mda G (2006) *Micronucleus test on gas station attendants.* Genet. Mol. Res. 5: 45–54.

Bishop J (1987) *The molecular genetics of cancer.* Science 235: 305-311.

Blakely RD, DeFelice LJ, and Galli A (2005) *Biogenic amine neurotransmitter transporters: just when you thought you knew them.* Physiol (Bethesda) 20: 225–231.

Blockwell RP, Maslanyj MP, Simpson J, Allen SG and Day NE (2002) *Exposure to power frequency electric field and risk of Childhood cancer in the UK.* Br. J. Cancer 87: 1257-1266.

Bohr H, Brunak S and Bohr J (1997) *Molecular wring resonance in chain molecules.* Bioelectromagnetics 18:187-189.

Boice JD Jr and Lubin JH (1997) *Occupational and environmental radiation and cancer.* Cancer Causes and Control 8: 309-322.

Bonassi S, Ugolini D, Kirsch-Volder M, Stromberg U, Vermeulen R and Kyrtopoulos JDSA (2006) *Biomarkers in environmental carcinogenesis research; striving for a new momentum.* Toxicol. Lett. 162: 3-15.

Cairns J (1975) *Mutational selection and the natural history of cancer.* Nature 255: 197-200.

Carborari K, Gonealves L, Roth D, Moreira P, Fernandez R and Martino-Roth MG (2005) *Increased micronucleated cell frequency related to exposure to radiation emitted by computer cathode ray tube video display monitors.* Genetics and Molecular Biology 28: 3, 469-474.

Cardis E, Richardson L, Deltour I, Armstrong B, Feychting M, Johansen C, Kilkenny M, McKinney P, Modan B, Sadetzki S, Schuz J, Swerdlow A, Vrijheid M, Auvinen A, Berg G, Blettner M, Bowman J, Brown J, Chetrit A, Christensen HC, Cook A, Hepworth S, Giles G, Hours M, Iavarone I, Jarus-Hakak A, Iaeboe L, Krewski D, Lagorio S, Lonn S, Mann S, McBride M, Muir K, Nadon L, Parent ME, Pearce N, Salminen T, Schoemaker M, Schlehofer B, Siemiatycki J, Taki M, Takebayashi T, Tynes T, Tongeren VM, Vecchia P, Wiart J, Woodward A and Yamaguchi N (2007) *The INTERPHONE study: design, epidemiological methods, and description of the study population.* Eur J Epidemiol. (Epub ahead of print – OPEN ACESS) http://www.springerlink.com/content/x88uu6q103076p53/

Casartelli G, Monteghirfo S, De Ferrari M, Bonatti S, Scala M, Toma S, Margarino G and Abbondandolo A (1997) *Staining of micronuclei in squamous epithelial cells of human oral mucosa.* Anal. Quant. Cytol. Histol. 19: 475-481.

Cerqueira EMM, Meireles JRC, Lopes MA, Junqueira VC, Gomes-Filho IS, Trindade S and Machado-Santelli GM (2008) *Genotoxic effects of X-ray on keratinized mucosa cells during panoramic dental radiography.* Dentomaxillofacial Radiology 37: 398-403.

Collins AR (1992) *Meeting Report: Workshop on "Single Cell Gel Electrophoresis (The Comet Assay)" held as part of the UKEMS/DNA repair Network Joint Meeting, Swansea.* Mutagenesis 7: 357-358.

Collins AR (2004) *The comet assay for DNA damage and repair: principals, applications, and limitations.* Mol. Biotechnol. 26: 249-261.

Collins AR, Duthie SJ and Dobson VL (1993) *Direct enzymatic detection of endogenous oxidative base damage in human lymphocyte DNA.* Carcinogenesis 14: 1733–1735.

Collins AR, Harrington V, Drew J, Melvin R (2003) *Nutritional modulation of DNA repair in a human intervention study.* Carcinogenesis 24: 511–515

Collins AR, Oscoz AA, Brunborg G, Gaivao I, Giovannelli L, Kruszewski M, Smith CC and Stetina R (2008) *The comet assay: topical issues.* Mutagenesis 23: 143-151.

Corvi R, Albertini S, Hartung T, Hoffman S, Maurici D, Pfuhler S, Benthem J and Vanparys P (2008) *ECVAM retrospective validation of in vitro micronucleus test (MNT).* Mutagenesis 23: 4, 271-283.

Davis UC (2006) *Radiation and human health.* Environment Health and Safety 2:1493 (http://ehs.ucdavis.edu/hs/dsa/index.cfm).

Diem E, Schwarz C, Adlkofer F, Jahn O and Rudiger H (2005) *Non-thermal DNA breakage by mobile-phone radiation (1800 MHz) in human fibroblast and in transformed GFSH-R17 rat granulose cells in vitro.* Mutat. Res. 583: 178-183.

Dillon VS, Thomas P and Fenech M (2004) *Comparison of DNA damage and repair following radiation challenge in buccal cells and lymphocyte using single cell gel electrophoresis.* Int. J. Radiat. Biol. 80: 517-528.

Dorr W, Dorschal B and Sprinz H (2000) *Report on the third annual meeting of the Society for Biological Research, GBS 99.* Radiat. Enviro. Biophys. 39: 147-152.

Duarte V, Muller JG and Burrow CJ (1999) *Insertion of dGMP and dAMP during in vitro DNA synthesis opposite an oxidized form of 7,8 -dihydro-8-oxyguanine.* Nucleic Acid Res. 27: 496-502.

Fairbairn DW, Olive PL and O'Neill KL (1995) *The comet assay : a comprehensive review.* Mutat. Res. 33: 37-59.

Fenech M and Crott JW (2002) *Micronuclei, nucleoplasmic bridges and nuclear buds induced in folic acid deficient human lymphocytes-evidence for breakage-fusion-bridge cycle in the cytokinesis-block micronucleus assay.* Mutat. Res. 504: 131-136.

Fenech M and Morley AA (1986) *Cytokinesis-block micronucleus method in human lymphocytes: effect of in vivo ageing and low dose X-irradiation.* Mutat. Res. 161: 193-198.

Fenech M, Chang WP, Kirsch-Volders M, Holland N, Bonassi S and Zeiger E (2003) *Human MicronNucleus project. HUMN project: detailed description of the scoring criteria for the cytokinesis-block micronucleus assay using isolated human lymphocyte cultures.* Mutat. Res. 534: 65–75.

Friedberg EC (2001) *How nucleotide excision repair protects against cancer.* Nat. Rev. Cancer 1: 22-33.

Gadhia PK (1998). *Possible age-dependent adaptive response to a low dose of x-rays in human lymphocytes.* Mutagenesis 13: 2151–152.

Gadhia PK, Shah T, Mistry A, Pithawala M and Tamakuwala D (2003) *A preliminary study to assess possible chromosomal damage among users of digital mobile phones.* Electromagn. Biol. Med. 22:159-169.

Garaj-Vrhovac V and Kopjar N (1998) *The comet assay- a new technique for detection of DNA damage in genetic toxicology studies and human biomonitoring.* Period. Biol. 100: 361-366.

Garaj-Vrhovac V, Fucic A and Horvat D (1992) *The correlation between the frequency of micronuclei and specific chromosome aberration in human lymphocytes exposed to microwave radiation in vitro.* Mutat. Res. 281: 181-186.

Glasstone S and Jordan WH (1981) *Nuclear power and its environmental effects.* American Nuclear Society. La Grange Park, Illionis.

Glaviano A, Mothersill C, Case CP, Rubio MA, Newson R and Lyng F (2009) *Effects of hTERT on genomic instability caused by either metal or radiation or combined exposure.* Mutagenesis 24: 1, 25-33.

Groves FD, Page WF, Gridley G, Lismeque L, Stewart PA and Tarone RE (2002) *Cancer in Korean war nevy technician:mortality survey after 40 years.* Am. J. Epidemiol. 155: 810-818.

Guelman LR, Zorrilla MAZ, Rios H, Di Toro CG, Dopico AM and Zieher LM (2001) *Motor, cytoarchitectural and biochemical assessment of pharmacological neuroprotection against CNS damage induced by neonatal ionizing radiation exposure.* Brain Res. Protoc. 7: 203-210.

Halliwell B (1996) *Free radicals proteins and DNA: oxidative damage versus redox regulation.* Biochem. Soc. Trans. 24: 1023-1027.

Hardell L, Lemart P, Anneli H, Arme M, Hansosson K and Carlberg M (2002) *Cellular and cordless phones and the risk of brain tumor.* Euro. J. Cancer Prev. 11: 377-386.

Hardell L, Mild KH and Carlberg M (2003) *Further aspects on cellular and cordless telephones and brain tumors.* Int. J. Oncol. 22: 2, 76-80.

Hardell L, Mild KH, Carlberg M and Soderqvist F (2006) *Tumour risk associated with use of cellular telephones or cordless desktop telephones.* World J. Surgical Oncology 4: 74-79.

Hermann DM and Hossmann KA (1997) *Neurological effects of microwave exposure related to mobile communication.* J. Neurological Science 152: 1-14.

Hocking B (1998) *Preliminary report: symptoms associated with mobile phone use.* Occup. Med. 48:6, 357-360.

Hruby R, Neubauer G, Kuster N and Frauscher M (2008) *Study on potential effects of "902-MHz GSM-type Wireless Communication Signals" on DMBA-induced mammary tumours in Sprague-Dawley rats.* Mutat. Res. 649: 34-44.

Huuskonen H, Lindbohm and Juutilainen J (1998) *Teretogenic and reproductive effects of low frequency magnetic fields.* Mutat. Res. 410:167-183.

IARC (2002) *Extreamely Low-Frequency Electric and magnetic Fields: IARC Monographs on the Evalution of carcinogenic Rick to Humans. Vol. 20.* Intrnational Agency for Research on Cancer, Lyon.

ICNIRP (2008) *ICNRP statement on EMF-emitting new technologies.* Health Physics 94: 4, 376-392.

ICNRP *(International Commission on Non-Ionizing Radiation Protection) (1998) Guidelines for limiting exposure to time-varying electric, magenatic Fields (up to 300 GHz).* Health Physics 74: 494-520.

Ingole IV and Ghosh SK (2006) *Cell phone radiation and developing tissue in chick embryo- a light microscopic study of kidneys.* J. Anat. Soc. India. 55: 2, 19-23.

Jiang H, Li W, Li X, Cai L and Wang G (2008) *Low-dose radiation induces adaptive response in normal cells, but not tumor cells: In vitro and in vivo studies.* J. Radiat. Res. 49: 219-230.

Johansen C, Boice JD, McLaughlin JJK and Olsen JH (2001) *Cellular telephones and cancer- a nationwide cohort study in Denmark.* J. National Cancer Institute 93: 203-207.

Joksic G, Petrovic S and Llie Z (2004) *Age related changes in radiation induced micronuclei among healthy adults.* Brazilian J. Medical and Biological Research 37: 1111-1117.

Juutilainen J and Liimatainen A (1986*) Mutation frequency in Salmonella exposed to week 100- Hz magnetic fields.* Heredites 104: 145-147.

Kadihim MA, Lorimore SA, Townsend KM, Goodhead DT, Buckle VJ and Wright EG (1995) *Radiation induce genomic instability: delayed cytogenetic aberration and apoptosis in primary human bone marrow cells.* International J. of Radiat. Biol. 67: 287-293.

Karinen A, Heinavaara S, Nylund R and Leszynski D (2008) *Mobile phone radiation might alter protein expressin in human skin.* BMC Genomics. 9: 77-79.

Kassam SN and Andrew JR (2009) *UV-inducible base excision repair of oxidative damaged DNA in human cells.* Mutagenesis 24: 1, 75-83.

Kawanishi S, Hiraku Y and Oikawa S (2001) *Mechanism of guanine-specific DNA damage by oxidative stress and its role in carcinogenesis and aging.* Mutat. Res. 488: 65-76.

Kirsch-Volders M, Mateuca RA, Roelants M, Tremp A, Zeigar E, Bonssi S, Holland N, Chang WP, Aka PV, De Boeck M, Godderis L, Haufroid V, Ishikawa H, Laffon B, Marcos R, Migliore L, Morppa H, Teixeira JP and Zijjno FA (2006) *The effects of GSTM 1 and GSTT 1 polymorphisms on micronucleus frequencies in human lymphocytes in vivo.* Cancer Epidemiol. Biomarkers Prev. 15: 1038-1042.

Kopjar N and Garaj-Vrhovac V (2001) *Application of the alkaline comet assay in human biomonitoring for genotoxicity: a study of Croatian medical personnel handling antineoplastic drugs.* Mutagenesis 16: 71-78.

Kovacic P and Jancitho JD (2001) *Mechanism of carcinogenesis: focus on oxidative stress and electron transfer.* Curr. Med. Chem. 8: 773-796.

Krewski D, Byus C, Glickman B, Lotz W, Mandeville R and McBride M (2001) *Recent advances in research on radiofrequency fields and health.* J Toxicol Environ Health B Crit Rev 4: 1, 145-159.

Kumaravel TS and Jha AN (2006) *Realible comet assay measurements for detecting DNA damage induced by ionizing radiation and chemicals.* Mutat. Res. 605: 7-16.

Kumaravel TS, Vilhar B, Faux P and Jha AN (2007) *Comet assay measurements: perspective.* Cell Biol. Toxicol. DOI 10.1007/s10565-007-9043-9.

Lahkola A, Auvinen A, Raitanen J, Schoemaker MJ, Christensen HC, Feychting M, Johansen C, Klaeboe L, Lonn S, Swerdlow AJ, Tynes T and Salminen T (2007) *Mobile phone use and risk of glioma in 5 North European countries.* Int. J. Cancer. 120: 8, 1769-1775.

Lai H and Singh NP (1995) *Acute low-intensity microwave exposure increases DNA single-strand breaks in rat brain cells.* Bioelectromagnetics 16: 3, 207-210.

Lai H and Singh NP (1996) *Single – and double-strand breaks in rat brain cells after acute exposure to radiofrequency electromagnetic radiation.* Int. J. Radiat. Biol. 69: 4, 513-521.

Lai H and Singh NP (1997) *Acute exposure to a 60 Hz magnetic field increases DNA strands break in rat brain cells.* Bioelectromagnetics 18: 156-165.

Lai H and Singh NP (2004) *Magnetic-field-induced DNA strand breaks in brain cells of the rat.* Environ. Health Perspect. 112: 6, 687-694.

Lantow M, Schuderer J, Hartwig C and Simko M (2006): *Free radical release and HSP70 expression on two human immune-relevant cell lines after exposure to 1800 MHz radiofrequency radiation.* Radiat Res. 165: 88-94.

Lavin MF, Khanna KK and Watters D (1999) *Sensing damage in DNA: ATM and cellular signaling.* Radiat. Res. 2: 348-351.

Lindahl T and Wood RD (1999) *Quality control of DNA repair.* Science 286: 1897-1905.

Lixia S, Ke Y, Kaijun W, Deqiang L, Huajun H, Xiangwei G, Baohong W, Wei Z, Jianling L and Wei W (2006) *Effects of 1.8 GHz radiofrequency field on DNA damage and expression of heat shock protein 70 in human lens epithelial cells.* Mutat. Res. 602: 135-142.

Lonn S and Ahloom A (2005) *Longe term mobile phone use and brain cancer risk.* Am. J. Epidem. 161: 526-535.

Loscher W and Liburdy RP (1998) *Animal and cellular studies on carcinogenic effects of low frequency (50/60 Hz) magnetic fields.* Mutat. Res. 410: 185-220.

Maes A, Collier M, Slaets D and Verschaeve L (1996) *954 MHz microwaves enhance the mutagenic properties of mitomycin-C, Environ.* Mol. Mutagen. 26-30.

Maish D (2003) *Chiildren and mobile phones, Is there a health risk?* J. Australasian College of Nutritional and Environmental Mediicine. 22: 2, 3-8.

Matsumoto H, Hamada N, Takahashi A, Kobayashi Y and Ohnishi T (2007) *Vanguards of paradigm shift in radiation biology: Radiation-induced adaptive and bystander responses.* J. Radiat. Res. 48: 97-106.

Matta CF and Burkhardt S (2003) *Health risk of cellular telephones: the myth and the reality,* OPHA 1-20.

Matthews S (2003) *An observational and modeling study of the atmospheric flow over Nauru.* Ph.D. thesis, School of Chemistry, Physics, and Earth Sciences, Flinders University.

McKelvey-Martin VJ, Green MHL, Schmezer P , Pool-Zobel BL, De Méo MP and Collins A (1993) *The single cell gel electrophoresis assay (comet assay): a European review.* Mutat. Res. 288: 47–63.

McMillan TJ, Tobi S, Mateos S and Lemon C (2001) *The use of DNA double-strand breaks quantification in radiotherapy.* Int. J. Radiation Oncology Biol. Phys. 49: 2, 373-377.

Miyakoshi J, Kittakawa K, Takebe H (1997) *Mutation induction by high density, 50-Hz magnetic fields in human meWo cells exposed iin the DNA synthesis phase.* Int. J. Radiat. Biol. 7: 75-79.

Miyakoshi J, Yamagish N, Ohtsu S, Mohri K, Takebe H (1996) *Increase in the hypoxanthine-guanine phosphoribosyl transferase gene mutations by exposure to high density 50 Hz magnetic field.* Mutat. Res. 349: 109-114.

Mohammadi S, Taghavi-Dehaghani M, Gharaati MR, Masoomi R and Ghiassi-Nejad M (2006) *Adaptive response of blood lymphocytes of inhabitants residing in high background radiation areas of Ramsar- micronuclei, apoptosis and comet assays.* J. Radiat. Res. 47: 279-285.

Mohseni-Meybodi A, Mozdarani H and Mozdarani S (2009) *DNA damage and repair of leukocyte from Fanconi anemia patients, carriers and healthy individuals as measured by the alkaline comet assay.* Mutagenesis 24: 1, 67-73.

Morgan RW, Kelsh MA, Zhao K, Exuzides KA, Heringer S and Negrete W (2000) *Radio frequency exposure and mortality from cancer of the brain and lymphatic/hematopoetic systems.* Epidemiology 11: 118-127.

Mortazavi SMJ, Ghiassi-Nejad M, Karam PA, Ikushima T and Niroomand-Rad A (2006) *Cancer incidence in area with elevated levels of natural radiation.* Int. J. Low Radiation 2: ½, 20-27.

Mothersill C and Seymour C (2006) *Radiation-induced bystander effects: do they provide evidence for an adaptive response?* Int. J. Low Radiation 2: ½, 119-127.

Mulder LC, Chakrabarti LA, Muesing MA (2002) *Interaction of HIV-1 integrase with DNA repairs protein hRad18.* J Biol Chem 277: 27489–27493.

Muscat JE and Malkin MG (2002) *Hand held cellular telephones and risk of acoustic neuroma.* Neurology 58: 1304-1306.

Nersesyan AK and Llin AI (2007) *The micronucleus assay in exfoliated human cells: a mini-review of papers from CIS.* Cytology and Genetics 41: 2, 115-124.

Nishioka H, Nishi K and Kyokane K (1981) *Human sliva inactivates mutagenecity of carcinogens.* Mutat. Res. 85: 323-333.

Norppa H and Falck GC (2003) *What do human micronuclei contain?* Mutagenesis 18: 221-233.

NRPB (2003) *Health effects from radiofrequency electromagnetic fields, Report of an Independent Advisory Group on Non-ionizing Group on Non-ionizing Radiations*, A document of NRPB. 14, No. 2.

Panagopoulos DJ, Chavdoula ED, Nezis IP and Margaritis LH (2007) *Cell death induced by GSM 900-MHz and DCS 1800-MHz mobile telephony radiation.* Mutat. Res. 626: 69-78.

Panagopoulos DJ, Karabarbounis A and Margaritis LH (2004) *Effects of GSM 900-MHz mobile phone radiations on reproductive capacity of Drosophilia melanogaster.* Electromagnetic Biology and Medicine 23: 1, 29-43.

Panagoupoulos DJ and Margaritis LH (2008) *Mobile telephony radiation effects on living organisms.* In: Harper AC and Buress RV, Mobile Telephones. 107-149.

Parry JM and Sorrs A (1993) *The detection and assessment of the aneugenic potential of environmental chemicals: The European Community Aneuploidy Project.* Mutat. Res. 287: 3-15.

Petkau A (1987) *Role of superoxide dismutase in modification of radiation injury.* Br. J. Cancer 8: 87-95.

Pinto D, Ceballos JM, Garcia G, Guzman P, Del Razo LM, Vera E, Gomez H, Garcia A and Gonsebatt ME (2000) *Increased cytogenetic damage in outdoor painters.* Mutat. Res. 467:2,105–111.

Piperakis SM (2008) *Comet assay: a brief history.* Cell Biol. Toxicol. Guest Editorial. DOI 10.1007/s10565-008-9081-y.

Pitozzi V, Pallotta S, Balzi M, Bucciolini M, Becciolini A, Dolara P and Giovannelli L (2006) *Calibration of the comet assay for the measurement of DNA damage in mammalian cells.* Free Radic. Res. 40:1149–1154.

Piyathilake CJ, Macaluso M, Hine RJ, Vinter DW, Richards EW and Krumdieck CL (1995) *Cigarette smoking, intracellular vitamin deficiency, and occurrence of micronuclei in epithelial cells of the buccal mucosa.* Cancer Epidemiol Biomarkers Prev. 4: 751–758.

Pogribny I, Koturbash I, Tryndyak V, Hudson D, Stevenson SML, Sedelnikova O, Bonner W and Kovalchuk O (2005) *Fractionated low-dose radiation exposure leeds to accumulation of DNA damage and profound alterations in DNA and histone methylation in the Murine thymus.* Mol. Cancer Res. 3: 10, 553-561.

Ramirez A and Saldanha PH (2002) *Micronucleus investigation of alcoholic patients with oral carcinomas.* Genet. Mol. Res. 1: 3, 246-260.

Ravazzani P (2008) *The interpretation of the results of the research on electromagnetic fields and health in Europe: the EC coordination action EMF-NET.* Ann. Telecommun. 63: 11-15.

Regina ML, Moros EG, Pickard WF, Straube WL, Baty J, Roti JLR (2003) *The effect of chronic exposure to 835.62 MHz. FDMA or 847.74 CDMA radiofrequency radiation on the incidence of spontaneous tumors in rats.* Radiat. Res. 160: 143-151.

Remondini D, Nylund R, Reivinen J, Poulletier de Gannes F, Veyret B, Lagroye I, Haro E, Trillo MA, Capri M, Frenceschi C, Schalatter K, Gminiski R, Fitzener R, Tauber R, Schuderer J, Kuster N, Leszczynski D, Bersani F and Maercker C (2006) *Gene expression changes in human cells after exposure to mobile phone microwaves.* Proteomics. 6: 17, 4745-4754.

Riley PA (1994) *Free radical in biology: oxidative stress and the effects of ionizing radiations.* Int. J. Radiat. Biol. 65: 27-33.

Rosenthal M and Obe G (1989) *Effects of 50-hertzs electromagnetic fields on proliferation and on chromosomal alteration in human peripheral lymphocytes untreated or pretreated with chemical mutagens.* Mutat. Res. 210: 329-335.

Rosin MP (1992) *The use of micronucleus test on exfoliated cells to identify Anti-clastogenic action in humans: a biological marker for the efficacy of chemopreventive agents.* Mutat. Res. 267: 265-276.

Roti JLR, Mallyapa RS, Bisht KS, Ahern EW, Moros EG, Pickard WF and Straube WL (2001) *Neoplastic transformation in C3H 10T1/2 cells after exposure to 835.62 MHz FDMA and 847.74 MHz CDMA radiations.* Radiat. Res. 155: 239-247.

Russian National Committee on Non-Ionizing Radiation (2008) *MICROWAVE NEWS. (RNCNIRP),* <www.pole.com.ru/news_en.htm>:

Sarto F, Finotto S, Giacomelli L, Mazzotti D, Tomanin R and Levis A (1987) *The micronucleus assay in exfoliated cells of the human buccal mucosa.* Mutagenesis 2: 11-17.

Savitz DA, Wachtel H, Barnes FA, John EM and Tvrdik JG (1988) *Case-control study of childhood cancer and exposure to 60-Hz magnetic fields.* Am. J. Epidemiol. 128: 21-38.

Scarfi MR, Fresegna AM, Villani P, Pinto R, Marino C, Sarti M, Altavista P, Sannino A and Lovisolo GA (2006) *Exposure to radio frequency radiation (900 MHz, GSM signal) dose not affect micronucleus frequency and cell proliferation in human peripheral blood lymphocytes: an interlaboratory study.* Radiat. Res. 165: 655-663.

Schmid W (1975) *The micronucleus test.* Mutat. Res. 31: 9-15.

Schoemaker MJ, Swerdlow AJ, Hepworth SJ, McKinney PA, Tongeren MV and Muir KR (2006) *History of allergies and risk of glioma in adults.* Int. J. Cancer 119:9, 2165-2172.

Singh NP, McCoy MT, Tice RR and Schneider LL (1988) *A simple technique for quantitation of low levels of DNA damage in individual cells.* Exp. Cell Res. 175: 184-191.

Singh NP, Stephens RE and Schneider EL (1994) *Modification of alkaline microgel electrophoresis for sensitive detection of DNA damage.* Int. J. Radiat. Bio. 66: 23-28.

Smith CC, Adkins DJ, Martin EA and Donovan MR (2008) *Recommendation for design of rat comet assay.* Mutagenesis 23: 3. 233-240.

Somosy Z (2000) *Radiation response of cell organelles.* Micron 31: 165-181.

Speit G and Schmid O (2006) *Local genotoxic effects of formaldehyde in human measured by the micronucleus test with exfoliated epithelial cells.* Mutat. Res. 613: 1-9.

Speit G, Schutz P and Hffmann N (2007) *Genotoxic effects of exposure to radiofrequency electromagnetic fields (RF-EMF) in cultured mammalian cells are not indepently reproducible.* Mutat. Res. 626: 42-47.

Stich HF and Rosin MP (1983) *Quantitating the synergistic effects of smoking and alcohol consumption with the micronucleus test on human buccal mucosa cells.* Int. J. Cancer 31: 305-308.

Stich HF, Rosin MP and Vallejera MO (1984) *Reduction with vitamin A and beta-carotene administration of proportion of micronucleated buccal mucosal cells in Asian betel nut and tobacco chewers.* Lancet 1204-1206.

Stich HF, Stich W and Parida BB (1982) *Elevated frequency of micronucleated cells in the buccal mucosa of individuals at high risk for oral cancer: Betel Quid chewers.* Cancer Lett. 17: 125-134.

Thomas P, Harvey S, Gruner T and Fenech M (2008) *The buccal cytome and micronucleus frequency is substantially altered in Down's syndrome and normal ageing compared to young healthy controls.* Mutat. Res. 638: 37-47.

Tice RR (1995) *The single cell gel/comet assay: a microgel electrophoretic technique for the detection of DNA damage and repair in individual cells.* In Phillips DH and Venitt, S (eds), Environmental Mutagenesis. Bios Scientific, Oxford, UK, 315–339.

Tice RR, Agurell E, Anderson D, Burlinson B, Hartmann A, Kobayashi H, Miyamae Y, Rojas E, Ryu JC, Sasaki YF (2000) *Single cell gel/ Comet Assay: Guidelines for in vitro and in vivo genetic toxicology testing.* Environ. Mol. Mutagen. 35: 206-221.

Tice RR, Andrew PW, Hirai O and Singh NP (1990) *The single cell gel (SCG) assay: an electrophoretic technique for detection of DNA damage in individual cells.* In Witmer CM, Snyder RR, Hollow DJ, Kalf GF, Kocsis JJ and Sipes JG (eds.), Biological reactive

Intermediates, IV. Molecular and cellular effects and their impact on human health. Plenum Press, New York, NY. 157-164.

Tice RR, Andrews PW, Hirai O and Singh NP (1991) *The single cell gel (SCG) assay: an electrophoretic technique for the detection of DNA damage in individual cells.* Adv. Exp. Med. Biol. 283:157-64.

Tolbert PE, Shy CM, Allen JW (1992) *Micronuclei and other nuclear anomalies in buccal smears: methods development.* Mutat. Res. 271(1):669-677.

TRAI (2008) Press Release No. 110/2008. www.trai.gov.in

Tynes T, Hannevik M, Andersen A, Vistenes Al and Haldorson T (1996) *Incidences of breast cancer in Norwegian female radio and telegraph operators.* Cancer Causes Control. 17:197-204.

Vainio H and Weiderpass E (2005) *From hazards identification to weighing the benefits and drawbacks of prevention.* Toxicol. Appl. Pharmacol. 207: 28-33.

Van-Deventer TE, Saunders R and Repacholi MH (2005) *WHO health risk assaement process for static fields.* Progress in Biophysics and molecular Biology 87: 355-363.

Vecchia P (2007) *Exposure of humans to electromagnetic fields.* Standards and regulations. Ann. Ist Super Sanita. 43: 3, 260-267.

Verschaeve L and Maes A (1998) *Genetic, carcinogenic and teratogenic effects of radiofrequency fields.* Mutat. Res. 410: 141-165.

Vijayalaxmi KC, Bisht WF, Pickard ML, Meltz JL, Roti EG and Moor A (2001) *Chromosome damage and micronucleus formation in human blood lymphocytes exposed in vitro to radiofrequency radiation at a cellular telephone frequency (847.7 MHz, CDMA).* Radiat. Res. 156: 430-432.

Vijyalaxmi BZ, Leal M, Szilagyi TJ, Pihoda ML and Meltz T (2000) *Primary DNA damage in human blood lymphocytes exposed in vitro to 2450 MHz radiofrequency radiation.* Radiat. Res. 153: 479-486.

Vilhar B (2004) *Help! There is a comet in my computer!* http://botanika.biologija.org/exp/comet/comet_guide01.pdf

Wang D, Kreutzer DA and Essigmann JM (1998) *Mutagenecity and repair of oxidative DNA damage: insight from studies using defined lesions.* Mutat. Res. 400: 99-115.

Weinberger Z and Richter ED (2002) *Cellular telephones and effects on the brain: The head as an antenna and brain tissue as a radio receiver.* Medical Hypotheses 59: 6, 703-705.

Wertheimer N and Leeper E (1979) *Electrical wiring configurations and childhood cancers.* Am. J. Epidemiol. 109: 273-284.

WHO (2000) *Electromagnetic fields and public health: mobile telephones and their base stations: fact sheet no.* 193 (June 2000 revision), World Health Organization, Geneva.

WHO (2005) *Children and mobile phones: Clarification statement* http://www.who.int/peh-emf/meetings/ottawa_june05/en/index4.html

WHO *Research Agenda for Radio Frequency Fields* (2006) http://www.who.int/peh-emf/research/rf_research_agenda_2006.pdf

WHO/*United Nation Environment Programme/International Radiation Protection Organization* (1984) Extremely low frequency (ELF) fields. Geneva: WHO; Environment Health Criteria 35.

WHO/*United Nation Environment Programme/International Radiation Protection Organization* (1987) Magnetic fields. Geneva: WHO; Environment Health Criteria 69.

WHO/*United Nation Environment Programme/International Radiation Protection Organization* (1993) Electromagnetic fields (300 Hz to 300 GHz). Geneva: WHO; Environment Health Criteria 137.

Williams JR, Zhang Y, Zhou H, Russell J, Gridley DS, Koch CJ and Little JB (2008) *Genotype-dependent radiosensitive: clonogenic survival, apoptosis and cell-cycle redistribution*. Int. J. Radiat. Biol. 84: 3, 151-164.

Yadav AS and Sharma MK (2008) *Increased frequency of micronuleated exfoliated cells among humans exposed in vivo to mobile telephone radiations*. Mutat. Res. 650: 175-180.

Yadav JS and Seth N (2000) *Effect of diagnostic X-rays on somatic chromosomes of occupationally exposed workers*. Ind J Expt Biol. 38: 46-50.

Yesilada E, Sahin I, Ozcan H, Yildirim H, Yologlu S and Taskapan C (2006) *Increased micronucleus frequencies in peripheral blood lymphocytes in women with polycystic ovary syndrome*. European J. of Endocrinology 154: 563-568.

Zeni O, Romano M, Perrotta A, Lioi AB, Barbieri R, D'Ambrosio G, Massa R and Scarfi M (2005) *Evaluation of genotoxic effects in human peripheral blood leucocytes following an acute in vitro exposure to 900 MHz radiofrequency fields*. Bioelectromagnetics 23: 258-265.

Zotti-Martelli L, Peccatori M, Maggini V, Ballardin M and Barale R (2005) *Individual responsiveness to induction of micronuclei in human lymphocytes after exposure in vitro to 1800- MHz microwave radiation*. Mutat. Res. 582: 42-52.

In: Mobile Phones: Technology, Networks and User Issues ISBN: 978-61209-247-8
Editor: Micaela C. Barnes et al., pp. 133-153 ©2011 Nova Science Publishers, Inc.

Chapter 4

DESIGNING MOBILE PHONE INTERFACES FOR COLLABORATIVE LEARNING IN EVERYDAY LIFE

Júlio Cesar dos Reis[1,2,*], Rodrigo Bonacin[1]
and Maria Cecília Martins[3,†]

[1]CTI Renato Archer, MCT, Rodovia Dom Pedro I, km 143,6 - Campinas, SP - Brazil
[2]Institute of Computing – University of Campinas (UNICAMP) – Campinas, SP – Brazil
[3]Nucleus of Informatics Applied to Education (NIED) – University of Campinas (UNICAMP) – Campinas, SP – Brazil

ABSTRACT

Literature points out new perspectives for collaborative learning through the use of mobile devices. This subject has being widely discussed by education and computer science researchers, showing as one of the strands for the future of (in)formal education. By using mobile devices, learning could be developed through collaborative interactions at any time or location. This represents a new possibility for people "learning while doing" their everyday activities. Moreover, the advent of mobile phones has created new opportunities that go beyond simple communication acts; their software interfaces have a primary role in enabling the collaboration among the evolved parties. We propose a novel approach which uses mobile collaborative learning for supporting everyday life tasks in general. In order to enrich the mobile collaborative learning, software applications should better explore the interfaces and multimedia resources available in current mobile devices. The mobile interfaces could support and boost situations that lead to learning, however it is essential to minimize the interaction difficulties, and maximize the learning activities itself. To achieve that, this work presents a design proposal and prototype of mobile phone interfaces for supporting mobile collaborative discussion, illustrating the ideas and the design decisions. This new approach aims to the enrichment of mobile interfaces, employing different medias and forms of interaction for the purpose of to constitute "wireless" communities of knowledge sharing about any issue or topic; thus stimulating and promoting the constitution of "communities of practice" through

---

[*] julio.reis@cti.gov.br and rodrigo.bonacin@cti.gov.br
[†] cmartins@unicamp.br

interaction, in which members can share common problems and/or work domain. In the proposed design, the resources of the interfaces are essentials to enable users to explain better their ideas; for that the paper presents multimedia interfaces to share images, sounds, and videos during the discussions. We also present a discussion about the impacts of this approach for informal education, and preliminary results from a qualitative analysis with real users.

Keywords: Mobile Collaborative Learning, Communities of Practice, Informal Education.

1. INTRODUCTION

Nowadays, mobile devices are present in a large range of human activities. The fast evolution of the technology, social network services and mobile platforms transformed the traditional notions of community and intercultural communication. Some researchers claim that a new connected and mobile society is emerging, with a variety of information sources and means of communication at home, at work, at schools and in the community [1]. These devices can help us to perform educational and leisure activities in a collaborative way, sharing knowledge of how to perform or perform them better. In this work, the informal education is understood as the lifelong education in which people learn from everyday experience, focusing in aspects related to our lives [2]. In the real world, most of the time we do not have lessons plans to follow, instead we respond to situations and experiences, as well as we learn from them.

In this scenario, mobility can be used for supporting education. Keegan [3] points out that the recent evolution in supporting technology for education can be seen firstly in distance education (d-Learning), then in electronic learning (e-Learning) and finally in mobile learning (m-Learning), called the "mobile revolution of the XXI century". Mobile learning can be developed by mobile devices such as: Personal Digital Assistant (PDA), Handhelds, Smartphones and Mobile phones. These devices can be exploited in the development of daily learning anywhere and/or anytime. According to Sharples [4], these mobile learning devices allow learners to learn wherever they are located and in their personal context then the learning is meaningful.

The mobile computing represents a new possibility for the people "learning while doing" their everyday activities. The advent of mobile devices has created new opportunities that go beyond the simple communication between people. There are new learning scenarios that can be supported by mobile devices, given that these devices can be present at anytime and anywhere. Therefore, by using mobile devices, it is possible to increase the educational development "out of school" by sharing knowledge and experiences about the situation that someone is living at the same time. However, subjects related to this type of education normally are diverse and comprehensive, consequently are necessary technological solutions that allow the involved people interact and act in a collaborative way.

There is a growing need to make m-Learning more interactive, stimulating the sharing of knowledge in a way that is not restricted to certain issues, place or time constraints. In order to achieve this objective, it is necessary to create situations that lead to the social and cultural impact on the use of mobile technologies, contributing in a practical way with the education and socialization of the citizens. At this context, interfaces have a primary role in enabling communication and collaboration among the evolved parties. In a learning environment for informal education, it is essential to design interfaces that minimize the interaction difficulties and maximize the learning activities itself. One big challenge is to deal with the devices restrictions, such as: screen size, performance, and data input difficulties. Another challenge is to provide appropriated interfaces to be used at different places, situations and contexts.

There are several relevant related works which have contributed with the mobile collaborative learning in the field of Mobile Computer Supported Collaborative Learning (MCSCL), (e.g. [5], [6], [7] and [8]). However, the approach adopted in these works focus mainly on classroom activities inside the school environment. Thus, it is also necessary to think and develop new appropriate methodologies and techniques for the use of mobile devices in learning processes outside the classroom. These methodologies and techniques can focus more than just in formal education, given their peculiar characteristics and capabilities that could support alternative forms of lifelong learning and informal education. Additionally, it is also necessary to have new technological solutions for mobile interfaces that allow the involved people to act in a collaborative way taking into account the self organization of the learning groups.

This work seeks for a solution that minimizes the mobile phone interaction limitations, and that can assist the development of people's daily informal education in a collaborative way. For that, we present a reflection on how the mobile collaborative learning could contribute to the development of informal education. Based on it a proposal is presented for a prototype of a mobile computing environment, illustring the design proposal of a mobile phone interface for mobile collaborative discussion. A new approach aims to the enrichment of this interface, employing different medias and forms of interaction. The objective is to stimulate the constitution of "communities of practice" in which members interact and share common problems and work in the same domain in order to constitute "wireless" communities of knowledge sharing about any issue or topic. Using multimedia interfaces is possible to practitioners share images, sounds, and videos during the discussions. In the proposed design, the resources of these interfaces are essentials to enable the practitioners to explain better their ideas. The proposed software application organizes the users' interaction, thus they can develop ideas and discussions collaboratively to solve problems in their daily lives. Therefore it is expected to create a technological solution that allows educational development in a mobile and collaborative way outside the school environment.

This work is organized as follows: Section 2 presents the theoretical referential; Section 3 presents a reflection on a new perspective for the development of informal education, explaining the approach and the design requirements, conception, principles and decisions; Section 4 presents the mobile software through the prototype and examples; Section 5 discusses the impact of the interface design decisions for the constitution of the communities and for the promotion of informal education; and Section 6 concludes and presents the future works.

2. THEORETICAL REFERENTIAL

This section presents the main theoretical background related to this proposal, discussing the informal education (section 2.1), mobile collaborative learning (section 2.2), and communities of practice (section 2.3).

2.1. Informal Education

According to Jeffs and Smith [9] the informal education brings back elements of an education dated to more than 2500 years ago. In ancient Greece, education was generally made on the streets at events in which people learned from each other through dialogues and discussions. Some of these educational characteristics and procedures are present in today's informal education. Fisher et al. [10] points out that informal education can be seen as the knowledge of the common sense, personal or practical knowledge, largely developed through experiences. For Smith [2] informal education does not have lessons or plans to follow. The informal is done through situations and experiences, and this can occur at any place, different from formal education, which is strongly linked to institutions and classrooms.

Besides, by not setting the time and location for the occurrence of activities, informal education is flexible for adapting the content of learning for each group in particular. Furthermore, Jeffs and Smith [9] clarify that one cannot say that informal education is better than formal education; it depends on the educational objectives, situations and of the context involved. It is also important to notice that according to Smith [2] the purpose of informal education is not different from any other form of education, it differs only in its scope and focus on aspects related to the common and everyday life. In addition to that, Glória [11] explains that informal education can be understood by multiple dimensions such as: (1) the learning and practice of exercises that enable individuals to organize into community goals, towards the solution of collective daily problems; (2) Learning the content of formal education in different forms and spaces with informal methodologies and (3) training of individuals to work through learning of skills and / or development of potential.

Informal education is related to a process of continuous learning, since we can learn all the time, every day and anywhere about a wide range of issues. In this sense, considering that this educational practice takes into account the learning that occurs on interactions and occupations emerged in the everyday life, it is possible to establish a relationship between informal education and mobile collaborative learning. This issue is explored in the next section.

2.2. Mobile Collaborative Learning

Learning can be seen like something socially built as the collaborative construction of knowledge. According to Dillenbourg [12], we cannot set a precise or exhaustive definition for collaborative learning. To sum up, it is a situation in which two or more people learn or attempt to learn something together interacting in a collaborative way. It describes a situation in which particular forms of interaction among people are expected to occur, which would trigger learning mechanisms. Hence, a general concern is to develop ways to increase the probability in which some types of interaction occur. Furthermore, collaborative learning must include situations, interactions, processes and effects.

Stahl et al. [13] point out that the collaborative learning involves individuals as group members, but also involves phenomena like the negotiation and sharing of meanings, including the construction and maintenance of shared conceptions of tasks, that are accomplished interactively in group processes. The basis of collaborative learning is in the interaction and exchange of information. Therefore technological mediums (hardware and software) that allow this interaction in an easier, simpler and more effective way can contribute to make this process more dynamic and effective. Collaborative learning through mobile devices has been investigated mainly because of the agility and mobility offered by these devices. Mobility has changed the contexts of learning and modes of collaboration, requiring different design approaches from those used in the traditional system developed to support teaching and learning. The major conclusion is that the learners' creations, actions, sharing of experiences and reflections are key factors to be considered when one is designing mobile collaborative activities for learning [14].

According to Roschelle et al. [15], MCSCL is a rapidly growing field with its intellectual activity focused on discovering, describing, and documenting the effectiveness of specific designs of use of mobile devices for learning in a collaborative way. These technologies provide new opportunities to promote and enhance collaboration by engaging learners in a variety of activities across different places and contexts. A main challenge is to identify how to design and deploy mobile tools and services that could be used to support collaboration in different kinds of settings. These different settings provide innovative ways for people and devices to interact by enabling learning to take place beyond the walls of the classroom and the screen of a computer [14]. Finally, Zurita and Nussbaum [7] clarify that the MCSCL activities support transparently the collaborative work by strengthening the: (a) organization of the managed material; (b) social negotiation space of group members; (c) enabling students to collaborate in groups by communication among the group members through the wireless network, that supports the social face-to-face network; (d) coordination between the activity states; (e) possibility to mediate the interactivity; (f) encouraging the mobility of members. Also, mobile collaborative learning activities manage and encourage tasks that include: monitoring real-time progress with respect to learning objectives and controlling interaction, negotiation, coordination and communication of the involved people.

The related works in the literature, which focus on design of mobile collaborative learning activities, (e.g., [7], [15], [8] and [14]) include: the proposition of methods and solutions that aims to solve questions for formal education environments inside the classroom; the investigation on how to design these applications with an interaction-based design [16]; the analysis of user's interaction [17]; and the approaches focused on specifics field and topics (e.g., [18] and [19]). The work of Breuer et al. [20] shows an approach to seamlessly integrate formal and informal learning, but the activities and the informal learning are still connected to the formal education inside the classroom. Other related works and approaches can be seen in [21], [22] and [23]. Nevertheless, none of these works have explicitly pointed out a particular solution to design a mobile application for the development of informal education outside the school environment. Therefore, this subject should be investigated in a deeper way, so it could create new educational paradigms not yet explored in the literature, and the proposal presented in this work is part of this scope of research.

2.3. Communities of Practice

Since the beginning of the history, human beings have formed communities that share cultural practices reflecting their collective learning: from a tribe around a cave fire to a group of nurses in a ward, to a street gang, or to a community of engineers interested in some issue. Participating in these "communities of practice" (CoP) is essential to the learning process. It is the core of what makes human beings capable of meaningful knowing [24].

The communities of practice are based in the social theory of learning. According to Wenger [25] this theory integrates the components: practice, meaning, identity and community as necessary to characterize social participation as a process of learning and of knowing. The main idea of the CoP is the individual as an active participant in the practices of social communities with common interest in some subject or problem, and that s(he) can collaborate and share ideas. These communities are in everywhere and people belong to a number of them: at work, at school, at home, and even in hobbies. "We are core members of some and we belong to others more peripherally. For example, you may be a member of a band, or you may just come to rehearsals to hang around with the group" [26].

Wenger [26] explains that members of a community are "informally bound by what they do together — from engaging in lunchtime discussions to solving difficult problems — and by what they have learned through their mutual engagement in these activities. A CoP is thus different from a community of interest or a geographical community, neither of which implies a shared practice." According to Wenger [26] a CoP defines itself along three dimensions: what it is about, how it functions and what capability it has produced. In the next section a reflection will be presented about a new perspective for the development of informal education through the use of mobile collaborative learning by CoPs.

3. MOBILITY AND COLLABORATION USING MULTIMEDIA: A NEW PERSPECTIVE FOR DEVELOPING INFORMAL EDUCATION

Some studies try to answer where the education actually happens. Bentley [27] addresses the "lifelong education and for life" arguing that educational development can occur at any location or time during a lifetime; in this sense, the educational process is related to a process of continuous learning. The use of mobile devices with appropriate software applications could support and intensify opportunities for learning since it can enable interaction anywhere and anytime; therefore it can be an option for the development of informal education.

People live in a process of continuous learning all the time and not only connected to certain places and/or institutions. This process is collective and involves mainly action, meaning, identity and social participation through communication, dialogue and collaboration. In this one, each individual must develop himself/herself by their contributions, with an active, engaged and practice participation, collaborating and sharing ideas about something through communities. Mobile technological mediums seem like an interesting way to develop this process, contributing to turn it more dynamic and effective through agility and mobility.

This work uses the informal education as focus, since in this form of education any issue can be discussed and explored by users. It also considers the mobile phone interfaces for the constitution of communities of practice aiming to promote the informal education. The use of mobile learning can intensify the chances of learning due to time and place flexibility, creating in this matter, novel possibilities for the development of the informal education. In this context, the mobile learning can boost the development of the informal education since it opens new possibilities for action and relationship of the individual with the world, and therefore allows interventions through interactions and collaboration. Moreover, individuals in face of a new situation or problem are supposed to act in a more agile and flexible way if they share knowledge with others. The joining of these two forms of learning (mobile and collaborative) can provide a special condition for the occurrence of informal education, since they bring peculiar characteristics that can be best exploited through a properly designed mobile computational environment.

In this context, users can take advantage of the environment through the free exploration of these ideas and doubts in the interaction with each other, allowing the generation of new knowledge, and contributing to their education and development as citizens. Thus, the computational proposal presented in this work foresees that the participants must have more possibilities and freedom to interact and propose collaborative discussions on topics related to the interests and practices of the groups.

3.1. Multimedia in the Collaborative Learning

The multimedia technology has transformed the ways that students communicate, learn and socialize themselves, improving their skills for presentation and exploitation of the knowledge. The use of multimedia must be more and better explored in educational contexts, especially in the informal ones, since it can contribute to a better and diversified formation of the individuals [28].

The multimedia instructional messages can be strongly used in the learning environments, since they are characterized by the exploitation of new knowledge and feelings. For Mayer [29] multimedia instructional messages is the communication using words, images (in movement or not) which intends to promote a learning. According to this principle, students learn better with words and pictures than with words alone. His studies indicate that when words and images are presented on a joint, students can build verbal and pictorial mental models, and also build relationships between these models. When words are presented alone, students just are able to develop verbal mental models, but they are less susceptible to develop a pictorial mental model and make connections between them.

Thus, various media should be available for the individual in order to provide different resources to express and make his/her reasoning understandable for the group. Collaboration between learners can be benefited from representations such as images and animations, as they may have the role of referential anchors in the building of a shared understanding. With the use of multimedia, users can employ some artifice like image, video or sound to explain their point of view quickly and easily in a context of collaborative learning [30].

The use of multimedia in a collaborative learning environment leads the user to ratiocinate about how to explain his/her thinking, (e.g. through an image); thereby creating means for the development of his reasoning. Furthermore, other users of the environment can use their imagination to understand the correlation of the image, video or sound along with text and the context under study, reaching conclusions on the explanation of the ideas from the involved. In this perspective, the approach to the exploration of multimedia in a mobile application to support the collaborative learning is plausible, since all necessary resources to capture and storage the audio-visual resources is already found in the mobile devices. The next section presents a mobile prototype interface designed based on the approach discussed in this section.

4. DESIGNING A MOBILE SOFTWARE APPLICATION FOR SUPPORTING EVERYDAY LEARNING: PROTOTYPE AND EXAMPLES

Due to the informal education heterogeneity, a simple and trivial computational solution (i.e. based only on exchange of asynchronous messages by instant text messages) is not able to support efficiently the collaborative situations. In order to build the prototype was necessary to develop and organize interfaces and different forms of interaction, with multiplies forms of collaboration that must be available in the environment. It was also necessary to provide the minimum of information that must be contained on these interfaces in a simple way. Several requirements was considered such as the flexibility of the application components to fit the different contexts of collaborative use (which are numerous when it deals with informal education); additionally, the application should provide many ways of expression. With this objective the prototype adopted various media as a form to develop a richer and fruitful collaboration.

Thus, the prototype designed in this work shows an organized set of design interfaces capable of supporting collaboration in synchronous and asynchronous ways. Additionally, we proposed a mechanism to consolidate (highlight) the messages in a synchronous collaborative process, and to vote the situations (states) of the developed collaborations. Figure 1 shows a general view of the proposed environment.

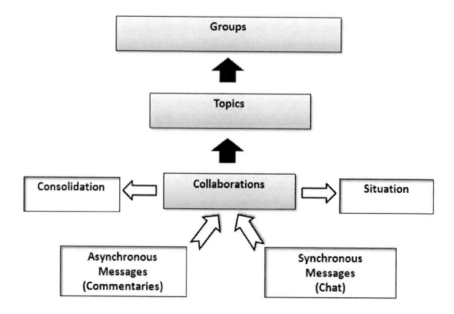

Figure 1. General view of the mobile collaborative environment.

According to the scheme presented in Figure 1, collaborations occur in the environment by sending synchronous (chat) and asynchronous messages (commentaries). The users collaborate to consolidate the collaboration, which gives emphasis to specific messages, and to provide the status of the collaborations. The computational proposal foresees that the participants must have more possibilities and freedom to interact and propose collaboratively discussions on topics related to the interests and practices of the groups through the software. The goal is to provide a virtual mobile environment that can support the constitution of CoP, in this way, individuals with similar interests and practices can constitute group of collaborators. These ones can discuss an existing topic and/or open new topics of discussion according to their interests of practice.

The communities are represented by groups in the software. They are created by users in order to organize topics related to the main group proposes. Furthermore, due to the large range of topics that can be discussed, it was created a way to organize the information in hierarchical levels: (1) Groups (communities), first division of themes; Topics, within a group there are several topics, and (3) Collaborations, numerous collaborations can be associated to one topic.

The process of developing collaboration can start by inserting a new group, or choosing a group that already exists. The groups represent the CoP and they have a vital role within the environment, since they organize the various themes that can be related to informal education into specific areas. Users of a same group are people with common interests and practices in which the objective is to discuss problems and find out solutions in a collaborative way. It means that a group is a specific area of formal or informal knowledge where users can be grouped, and topics and collaborations are organized. Users are able to join the existing groups, or create new ones, as they want. In the next section, the developed prototype will be presented through examples.

4.1. Prototype and Examples

The main principle behind the proposed design is the self organization of the groups without any formal moderation. These groups perform discussions through collaboration sessions. In the prototype, the collaboration occurs through synchronous and asynchronous multimedia messages. A specific group of collaborators, in a discussion, can consolidate (highlight) messages that could be important to someone that may want to know a synthesis of it. The users can decide which state (situation) each collaboration is, for example any collaborator could vote if they find out the solution of a problem or not. The users can also search discussions by topic in order to know more about some problem, and give opinions (commentaries) using asynchronous messages even after the end of a synchronous session.

The development of the collaboration can start in the insertion of a new group (see Figure 2a and 2b). Within the groups can be added various topics (see Figure 3a and 3b) and within the topics is possible to add new collaborations. Figure 3b shows the list of topics for a group called Bikes, in this interface users can choose a topic to add a new collaboration to interact through synchronous and asynchronous messages.

 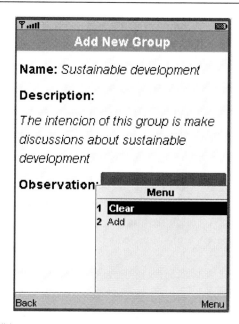

(a) (b)

Figure 2. Prototype's interfaces about groups.

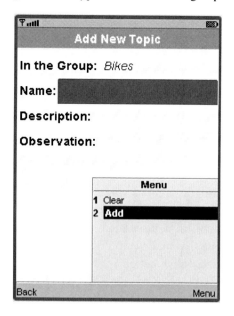 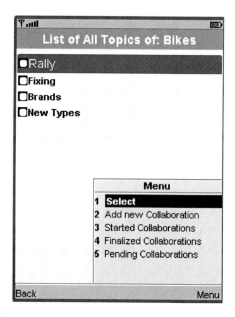

(a) (b)

Figure 3. Prototype's interfaces about topics.

During the collaborations participants can send synchronous messages as in an instant chat (Figure 4a illustrates this situation). All the messages are sent with a specific "objective" and have a label as: 'doubt', 'question', 'conclusion', 'answer', among others that may be defined by the own users following their needs of expression. After sending the message, it is displayed in an interface that centralizes the messages from all involved collaborators in the discussion, as shown in Figure 4b. This illustrates the interface with an example of synchronous message exchange in which users establish a communication (chat) from a defined theme – in this case "Nature and Profits" - and specify his "speech" as the type: question, answer, solution, questions, etc.

The participants can also select which synchronous messages should be "consolidated" during the chat (Figure 4c). These consolidated messages have an important role because they will describe a summary of all the synchronous interaction with the most important messages selected by users in a specific issue. At the consolidation interface, the messages are organized by their types (labels), for example, 'doubt', 'question', and 'conclusion'. Figure 4c shows an example of consolidated messages of the collaboration developed in Figure 4b. Other users of the environment can also see this summary (Figure 4c illustrates this interface).

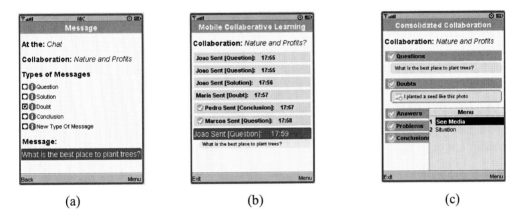

Figure 4. Prototype's interfaces about synchronous collaboration and consolidation of messages.

The occurrence of communication via asynchronous messages - as shown in Figure 5a - is a way to add new information to synchronous collaboration developed or under development. This type of communication will be especially useful in two situations:

1) When the theme of collaboration takes several days to be resolved, in which there is a need for several rounds of synchronous interaction (online chat). In this case the commentaries (asynchronous messages) can be a way to divulge possible solutions at any time between the online conversations. These commentaries can be discussed in a new round of synchronous collaboration, thus the discussions of the chat (synchronous messages) are articulated with the commentaries (asynchronous messages).
2) In case of a relevant idea of a user who was not involved in the collaboration after it has finished, the commentaries are a way for the users to register their idea so that other people can see.

As illustrated in Figure 5b, another proposed way for supporting the collaborative environment is to define possible "states" for the developed collaborations. Collaborators can vote on the basis of information from the collaboration (synchronous and asynchronous messages) to classify the status of the collaboration; e.g.: 'resolved', 'pending', 'not 'conclusive', among others that may be defined by users. This is an important functionality because this interface takes into account the opinions from users.

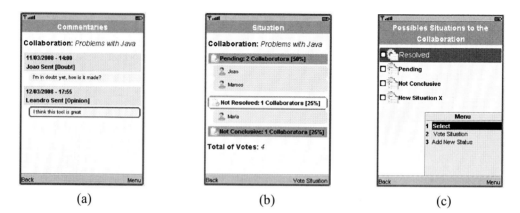

Figure 5. Prototype's interfaces about asynchronous collaboration (commentaries) and situation of the collaborations.

This state or situation refers to what has been discussed in a collaboration session (determined subject proposed by one person), so a situation would be, for example, the impossibility of conclusion or resolution. That is, the participants of the collaboration interacted with each other and exchanged messages, but they did not reach a definitive conclusion about the problem in question. Different final situations to a collaboration session can be defined in the prototype. The objective is to have a number of situations selected by the users of the application, presenting in this matter different points of view from the same collaboration, enabling users to check if that collaboration has generated interesting results or not (see Figure 5c).

Figure 4b illustrates the interface with an example of synchronous message exchange in which users establish a communication (chat) from a defined issue – in this case "Nature and Profits" - and specify his/her "speech" as being of the type: 'question', 'answer', 'solution', 'doubt', etc. Figure 4c shows an example of consolidated messages from the collaboration developed in Figure 4b. Figure 5a illustrates an interface of commentaries, in which users sent asynchronous messages to a collaboration named "Problems with Java". Finally Figure 5b shows the vote results made by users for the collaboration entitled "Problems with Java".

Besides these aspects, the environment encourages autonomy by providing resources for the self-organization through the collaboration sessions. Users can launch themes for discussion; they can be involved in the resolution of a problem, choosing the relevant solutions, and they can point out the status of the collaborations. Thus, the environment does not foresee the existence of a group's mediator, hence mechanisms have been developed in order to enable all users to organize the proposed subjects by the creation of groups, topics and collaborations necessary for the organization of the information at the mobile software.

4.2. Using Multimedia in the Mobile Collaborative Learning

The prototype developed use multimedia resources to explain ideas, facts or concepts using the media (audio, image and video) as a way for explanation and expression of ideas. Therefore the proposal is not to provide functionalities in a software environment to produce video or sound collaboratively, but to use these resources. There would have proposed means to make the explanation of the exact through the abstract, the art through images, audio and videos helping in a process of collaborative learning between users through mobile devices.

Consequently is discussed how the multimedia was introduced in the collaborative learning through the interfaces of the prototype. We should clarify that was assumed that the media archives were already recorded in the respective mobile devices, and the software interfaces only use the media that is already available.

In the prototype, besides other interfaces for the management of the collaborative environment, from the specific interface for sending messages, we developed an interface mechanism for the inclusion of multimedia; which provides means to include medias of the following types: image, audio and video. Thus, when users are interacting in the collaboration, they may attach media in their messages. Figure 6(a) shows an example of interaction (sending the message) using multimedia; for that, the user will need to click on the menu "Add media" in order to add a media in the message using the interface of messages.

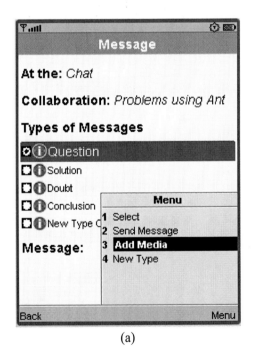 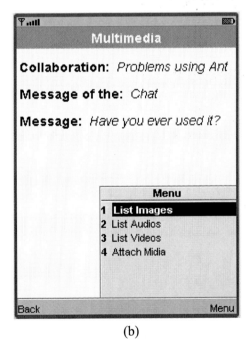

(a)   (b)

Figure 6. Prototype's Interface of message showing the option "Add Media" at the menu (a); Interface of Multimedia (b).

In sequence as shown by Figure 6(b), the multimedia interface appears to the user. In this interface users can see the name of the collaboration, the text message and what type of collaboration (chat or commentary) media will be attached to this message. At this moment, the user has the option to list the three types of media available to be manipulated in the environment, as shown the right side menu of Figure 6(b). People can choose a media file, as illustrated in Figure 7(a). This interface displays the list of all media available in the device, given the type chosen in the menu. After that users can visualize it, as shown in Figure 7(b) (the example shows an audio) that presents the media after the selection. In this way, users can attach the media in the message through an option from the menu in the interface shown in Figure 7(b).

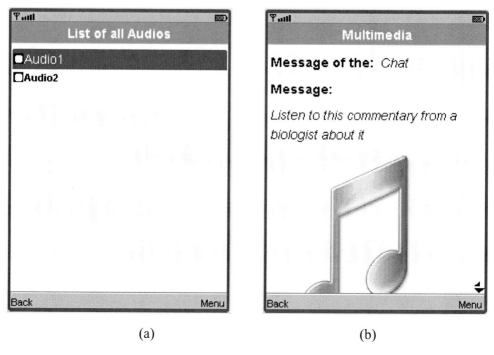

(a) (b)

Figure 7. Prototype's Interface of listing all audio files (a) and the media (sound) already attached to the text message (b).

After sending a new message to the collaboration, the media attached to the message will be available to other users. The sent message will have an icon on the interface of the collaboration that will indicate whether a media has been attached and what type of media. Figure 8 illustrates the types of icons that will appear in the interface indicating the media added.

(a) (b) (c)

Figure 8. Types of icons used to represent the several media available by the interface: (a) image, (b) audio and (c) video.

The icon in Figure 8(a) indicates that a picture was attached to that message; the icon in Figure 8(b) indicates that an audio was added to the message, and the icon in Figure 8(c) indicates that a video was attached to that message. Figure 9(a) illustrates the collaborative interface (chat) with the messages and the media attached to them. This resource in the interface indicates if there is media attached to a message or not. In order to view the content of this media, users must select the message, access at the right side the menu option "See media". Then the media will be displayed as illustrated in Figure 9(b) (interface that shows the media attached to a message). Note that with the media, users will see the text message related to that media; He can select the option "Back" at the left side menu to return to the interface with the collaborative messages (Figure 9a). Every message of the collaborative context (synchronous or asynchronous) sent by a user in the environment may have attached to an image, an audio or a video together with or without text. Figure 9(c) illustrates the interface of commentaries about a collaboration, presenting messages with attached media, as also happens in the chat (Figure 9a).

(a)   (b)   (c)

Figure 9. Prototype Interface showing the messages with media from users of the chat (a); Interface multimedia showing an image attached with the message of the chat in a collaboration called "Nature and Profits" (b); an Interface of Commentaries showing also media attached to the messages of this kind of collaboration (c).

It is important to note that an interesting way to introduce the multimedia in the collaboration is through the attachment of the media in the mobile collaboration messages. Given that the volume of messages that may arise during the collaboration sessions can be big, and also considering the screen size of mobile devices would be impracticable put all these information together in a single interface for manipulating media. Finally, after having presented a proposal on how the multimedia resources could be added to messages in an environment for supporting of mobile collaborative learning it is important to make a discussion about what this approach can educationally provide for users. Thus, the next section discusses this use and explains on some potential educational advantages that this environment can bring for people.

5. INTERFACES, COMMUNITIES AND INFORMAL EDUCATION: DISCUSSING THE PROPOSED APPROACH

Features designed for the environment must be represented on its interface, and it is necessary that interfaces make sense to users in their context. Furthermore, the design of the features, the interaction model and its relationship with the educational activities must be well designed. At learning in a virtual environment, the interfaces can be as a facilitator or a big problem, since if this interface is bad designed can let the educational activities unviable; however in the other way it can provide great support and really maximize the learning results.

Moreover, to provide the development of the communities, the interface should present specific features for satisfactorily support and management. The interface should provide mechanism for users create, organize and maintain the communities by themselves. The prototype developed in this work shows a possible alternative to stimulate the constitution of the CoP through groups. From these groups, all the organization of the application, the topics and the collaborative discussions are developed.

The themes, topics and collaborations supported by the application's interface create conditions for users develop the communities in a natural way. The subjects related to their daily life fits in the thematic of these communities, and the communities are the basis for the organization of themes that are part of the informal education. The design of the interface allows a particular organization which enables the establishment of these communities and therefore also issues from the informal education, which is developed through collaborative discussions.

In the proposed approach agility and mobility are viable through mobile devices which promote opportunities for situated and collaborative learning that occurs throughout life, anytime and anywhere. Learning experiences are encouraged in a process of communication and collaboration permeated by issues or questions related to situations experienced in people's lives. Besides the prototype allows people to build CoPs which can discuss problems in a little structured way, with the freedom to propose questions and solutions, to interact and express with decisions and solutions built through a collective consensus whatever the issue.

The mobile software was not thought to a specific application or to support a specific theme. A lot of real-world educational applications could be instantiate to any area. As an example, we could think on healthcare professionals that take care of people at home and need fast answer to problem and they do not have access to computer at that time. They could create a 'community' using this software to collaborate and to exchange ideas about problems that could appear on time of working and be resolved quickly using mobile collaboration. Other example in another context, users could collaborate to solve problems regarding to how to fix mechanical problem of car engines.

The use of multimedia resources in the interface adds distinct features to the application, providing new forms of expression available to the users during the collaboration. These resources can provide benefits to the exploration of the creativity of those involved, since they can make associations between the messages in the collaboration with the external world, such as by creating a video to illustrate an idea in a message, the user will be exercising his/her reflection and trying to make connections of that media with all the collaborative context under discussion and with the message that will be sent.

Other initiatives has explored how to provide and share media resources in a group of users [31, 8], however they does not explicitly use the audio-visual resources inside the collaboration. In this work, we highlight the importance of exploiting the advantages of combining multiple media in the computing environment. Otherwise, in this work was studied the use of these resources for supporting the explanation of ideas in the collaborative learning, such using it as multiple ways to explain a thought or idea in the collaboration. This approach may generate suitable means for learning be explored in a more incisive way on users' daily lives. The presented prototype can take advantage of the locations (places) where users are for the collaboration and educational development of the involved group. Including visuals resources in the collaborative mobile application users with mobile device can take advantages of the environment around them looking for explanations that exemplify their ideas in a clear and simple way.

The proposed solution allows the possibility for users to make connections between the collaborations under discussion in the software environment with their everyday life. Users may have difficult situations during the collaboration that only the use of simple words and phrases would not be enough to express an idea or an intention, thus multimedia resources could be used as alternative methods to solve this problem. Furthermore, the use of multimedia provides a richer learning environment in that users will have an alternative to words to express themselves. It is understood that using multimedia in a propitious environment to capture, to explore and to aggregate to a context or subject can generate many reflections by users, both in the development of the visual or audio "product" and in its relationship with the collaborative context which the media should be inserted. Users should think what media to use, and how to use it in order to their ideas be understood in a practice, clear and dynamic way by other users of the collaborative environment with this form of media (expression).

Concerning the empirical analysis, a brief qualitative evaluation of the proposal and prototype with potential target users has been done. At this moment the goal was not to verify issues of device connectivity and location aware information; however we intended to analyse characteristics and functionalities of the software.

This analysis has been conducted with participation of Information Technology professionals (4 participants), Computer Science students (3 participants), Healthcare professionals (2 participants) and high school students (2 participants). The participants have had contact with the prototype and answered the questions about their impression over the proposal. The questions elaborated were:

- If you face a problem, would you use the proposed application to discuss the solution? If yes, in which situation;
- What topics would you be more inclined to discuss?
- Would you collaborate with other people through the application to solve a problem proposed by someone? If yes, in which situation;
- Would you use multimedia messages during the collaboration?
- What difficulties could you point out in using the application in your daily life?
- How long and how often would you use it in your daily life?
- Do you think that you could improve (learning to develop better) the performance of your tasks with this application?

After analyzing the answers, we can indicate that users mainly would use the application in emergency situations, for example, when (s)he had no access to a computer, and also in working field situations or in occasions of daily life like such as during traffic time. They could discuss about several topics such as news, health, hobbies, technology, or on topics of everyday and professional context, generally in situations that need help. Most of users indicated that they would collaborate with other users on the application, especially in areas that they have knowledge. About the multimedia messages, all participants pointed out that they would use some form of multimedia messages, mainly due to the flexibility offered by these messages. The difficulties that could prevent users to use the application are mainly due to: the small screen of mobile devices, the difficulty of interaction (data entry), operation cost and low speed connection. Additionally, users have indicated that they would use the application especially when they had to solve a problem in unusual places and times, but they would not frequently use the application, as they could use personal computers in normal situations. Most users clarified in a positive way that a proposal with this approach could improve their performance in development of learning activities related to informal education. Looking to these results, we think that with this approach we can improve both tangible and intangible educational benefits for users, since they will be actively collaborating and learning without time, place and issue constraints.

Finally the use of the collaboration and mobility for the development of the informal education seems like a valid educational approach. It enables users to begin discussions about what they are experiencing in one moment about any issue, and also to participate in existing groups to share ideas and opinions. Then users can collaborate on a joint process of learning, in which the goal is mainly the intellectual evolution of the involved collectivity. To sum up, users could benefit from collaboration and flexibility through the use of mobile devices.

CONCLUSION

Recent studies have investigated how education can be boosted and developed in any place or time. They have investigated mainly how the learning process could occur regardless of subject or any other constraint aiming to maximize the possibilities for learning of the citizens in their daily lives. Mobile collaborative learning environments are frontiers of research in the scientific community of the area and there are several challenges in this context to obtain improvements in the quality of the collaboration. In order to achieve improvements on this aspect and in the users' experience in learning activities it is essential to study new alternatives to design richer mobile computing environments to improve the capacity and easiness of communication, interaction, and expression. For that, this work presented a technological proposal for the development of the informal education based on mobile and collaborative learning, including a reflection about this new perspective. Based on it, a prototype was implemented aiming to illustrate the proposed approach for the design of mobile software, including the use of multimedia in mobile collaborative learning.

We addressed aspects for designing interfaces of a computational environment to support mobile collaborative learning, which allows the constitution of CoPs. These CoPs are developed from groups and mobile collaborative discussions, which aim to promote and develop the informal education. The developed prototype illustrated the main ideas of the approach. Using this prototype, an analysis about the role and impact of the mobile software interfaces was conducted, and we could observe that the proposal shows propitious scenarios for the development of the informal education constituting communities of practice. Moreover, we intended to perform an investigation on how audio-visual resources could be aggregated in the process. The approach shows as an interesting alternative to deal with the mobile interaction and collaboration restrictions; the prototype presented possibilities for exploiting in this direction. The features designed assist mainly the development of collaborative activities using the multimedia resources already available by the devices. They presented an interesting way to improve the learning through multimedia resources in a mobile application.

Based on the results obtained with potential users, the proposal can be pointed out as a likely starting point for the development of the informal education using mobile collaborative learning. It aims to add educational value for those involved at any place or time in a collaborative session. Moreover, this work emphasizes flexible ways to develop a collaboration session in a mobile phone application exploring the autonomy of the involved users with focus on informal education.

As further work is proposed a better and deeper investigation of the approach presented from a theoretical and practical point of view. Even though the proposal is based on these aspects, research on real case studies should point out improvements and new solutions for the design of the interface in the studied context. Concerning the features of the prototype, improvements in the design should be made; including those issues related to usability and to include new features in the prototype, mainly to sophisticate the establishment of the communities of practice. We propose to study the real educational benefits of the proposed approach in long period of use. Actually the use on a large scale should also provide what the concrete educational results of the proposed approach are.

REFERENCES

[1] Naismith, L., Lonsdale, P., Vavoula, G., Sharples, M. (2006). *Report 11: Literature Review in Mobile Technologies and Learning.* In: Future Lab. University of Birmingham, http://www.futurelab.org.uk/resources/documents/lit_reviews/Mobile_Review.pdf

[2] Smith, M. K. (1997*). Introducing informal education: What is informal education? Where does it happen? How has it developed?* http://www.infed.org/i-intro.htm#what

[3] Keegan, D. (2002). *The Future of Learning: From e-learning to m-learning.* (In Hagen Zentrales Institute fur Fernstudienforschung: Fern Universitat).

[4] Sharples, M. (2000). *The design of personal mobile technologies for lifelong learning.* Computers and Education, 34, 177–193.

[5] Roschelle, J. and Pea, R. (2002). *A walk on the wild side: How wireless handhelds may change CSCL.* In G. Stahl (Ed) Proceedings of CSCL 2002 (pp. 51-60). Boulder, CO.

[6] Zurita, G., Nussbaum, M., and Sharples, M. (2003). *Encouraging face-to-face collaborative learning through the use of handheld computers in the classroom.* In: Mobile HCI 2003 Fifth International Symposium on Human Computer Interaction with Mobile Devices and Services (pp. 193-208). Udine, Italy.

[7] Zurita, G., and Nussbaum, M. (2004). *Mobile Computer supported collaborative learning using wirelessly interconnected handheld computers.* Computers and Education, 42(3), 289-314.

[8] Arrigo M., Giuseppe, O., Fulantelli, G., Gentile, M., Novara G., Seta, L. and Taibi, D. (2007). *A Collaborative Mlearning Environment.* In Proceedings of the 6th International Conference on Mobile Learning (MLearn 2007) (pp. 13-21). Melbourn, Australia.

[9] Jeffs, T. and Smith, M. K. (1996). *Informal Education - conversation, democracy and learning.* (London: Educational Heretics Press).

[10] Fisher, T., Higgings, C. and Loveless, A. (2006). *A. Teachers Learning with Digital Technologies: A review of research and projects.* Retrieved February 03, 2009 from www.futurelab.org.uk/research/lit_reviews.htm

[11] Glória, M. (2005). *Non-Formal Education: Fields and problems.* Retrieved April 01, 2009 from http://www.kinderland.com.br/dwp_publicacoes.asp?id_sec=40

[12] Dillenbourg P. (1999). *What do you mean by collaborative learning?* (In P. Dillenbourg (Ed) Collaborative-learning: Cognitive and Computational Approaches. (pp.1-19). Oxford: Elsevier)

[13] Stahl, G., Koschmann, T., and Suthers, D. (2006). *Computer-supported collaborative learning: An historical perspective.* In R. K. Sawyer (Ed.) Cambridge handbook of the learning sciences (pp. 409-426). Cambridge, UK: Cambridge University Press.

[14] Spikol, D. (2008). *Playing and Learning Across Location: Identifying Factors for the Design of Collaborative Mobile Learning.* In School of Mathematics and Systems Engineering, Växjö University. Reports from MSI. Licentiate Thesis

[15] Roschelle, J., Rosas, R., and Nussbaum, M. (2005). *Towards a design framework for mobile computer supported collaborative learning.* In: International Society of the Learning Sciences Proceedings of the 2005 Conference on Computer support for Collaborative Learning: Learning 2005: the Next 10 Years! (pp. 520 - 524). Taipei, Taiwan.

[16] Lagos, M., Alarcón, R., Nussabaum, M., Capponi, F. (2007). *Interaction-Based Design for Mobile Collaborative-Learning Software.* Software, IEEE. 24(4), 80-89.

[17] Kong, S. (2008). *Collaborative Learning in a Mobile Technology Supported Environment: A Case Study on Analyzing the Interactions.* In Fifth IEEE International Conference on Wireless, Mobile and Ubiquitous Technology in Education (WMUTE 2008). (pp. 167-169). Beijing, China.

[18] Cabrera, J. S., Frutos, H. M., Stoica, A. G., Avouris, N., Dimitriadis, Y., Fiotakis, G. and Liveri, K. D. (2005). *Mystery in the museum: collaborative learning activities using handheld devices.* In ACM International Conference Proceeding Series Proceedings of the 7th international conference on Human computer interaction with Mobile Devices and Services (pp. 315 – 318). Salzburg, Áustria.

[19] Liu, C. C., Tao, S. Y., Ho, K. W., Liu, B. J., Hsu, C. C. (2007*). Constructing an MCSCL Groupware to Improve the Problem-solving Experience of Mathematics for Hearing-impaired Students.* In Seventh IEEE International Conference on Advanced Learning Technologies (ICALT 2007) (pp. 345-347).Niigata, Japan.

[20] Breuer, H., Konow, R., Baloian, N., Zurita, G. (2007). *Mobile Computing to Seamlessly Integrate Formal and Informal Learning.* In Seventh IEEE International Conference on Advanced Learning Technologies (ICALT 2007) (pp. 589-591). Niigata, Japan.

[21] Black, J. T. and Hawkes L. W. (2006). *A prototype interface for collaborative mobile learning. In International Conference on Communications and Mobile Computing.* In Proceedings of the 2006 international conference on Wireless communications and mobile computing (pp. 1277 – 1282). Vancouver, British Columbia, Canada.

[22] Lin, C. (2008). *A system perspective to establish a mobile collaborative learning environment (MCLE) - A preliminary study of Empirical Practice.* In: Fifth IEEE

International Conference on Wireless, Mobile and Ubiquitous Technology in Education (WMUTE 2008). (pp. 202-204). Beijing, China.

[23] Peter, Y., Vantroys, T., Leprête, E. (2008). *Enabling Mobile Collaborative Learning through Multichannel Interactions*. In: 4th International Conference on Interactive Mobile and Computer Aided Learning IMCL2008 (pp. 1-4). Amman, Jordan.

[24] Nicolini, D., Gherardi, S., Yanow, D. (2003). *Knowing in Organizations: A Practice-Based Approach*. M. E. Sharpe

[25] Wenger, E. (1999). *Communities of Practice: Learning, Meaning, and Identity*. Cambridge: Cambridge University Press

[26] Wenger, E. (1998). *Communities of Practice: Learning as a Social System*: In: Systems Thinker, http://www.co-i-l.com/coil/knowledge-garden/cop/lss.shtml

[27] Bentley, T. (1998). *Learning beyond the Classroom*: Education for a changing world. (London: Routledge).

[28] Anderer, C; Neff, J. M.; Hyde, P. (2007). *Multimedia Magic: Moving Beyond Text*. User Services Conference. Proceedings of the 35th annual ACM SIGUCCS conference on User services. Orlando, Florida, USA, p. 1 – 3.

[29] Mayer, R. E. (2001). *Multimedia Learning*. Cambridge University Press, Cambridge.

[30] Roschelle, J. (1992). *Learning by collaborating: Convergent conceptual change*. The Journal of the Sciences, 2, 235-276.

[31] Hwang, W. Y.; Hsu, J. L.; Huang, H. J. (2007). *A study on ubiquitous computer supported collaborative learning with hybrid mobile discussion forum*. Conference Proceeding of 6th Annual International Conference on Mobile Learning - 16–19 October 2007, Melbourne - Australia, p. 13.

Chapter 5

INTERNATIONAL EXPANSION OF EUROPEAN OPERATORS: A DESCRIPTIVE STUDY

Lucio Fuentelsaz[*], *Elisabet Garrido and Juan Pablo Maicas*
Universidad de Zaragoza, Zaragoza, Spain

ABSTRACT

This chapter describes the internationalization process followed by the main European mobile operators from 1998 to 2008. We measure the degree of internationalization by the number of countries in which they operate in European OECD countries and the rest of the world. In this chapter, we relate the international diversification to (i) the history of wireless communications in Europe, (ii) the evolution in the mode of market entry and, finally, (iii) the rebranding process followed by international wireless groups. This chapter aims to serve as a guide to the different patterns of internationalization of the main European operators.

INTRODUCTION

European mobile service providers experienced an important international growth during the last two decades (Curwen and Whalley, 2008). They started operating exclusively in their respective local markets at the beginning of the 90's and now we can observe global operators competing all over the world. If we consider the historical evolution of the industry, this international growth has been related to the technological evolution, especially as a consequence of the standardization of the wireless technology (Gruber and Verboven, 2001). The introduction of the digital GSM standard in Europe in the early 90's allowed a more efficient use of the radio spectrum and the subsequent increase in the number of players in all markets at the beginning of the 90's, when wireless communications were not so widespread among population. At that time, firms competed to attract users to their networks by decreasing prices and improving service quality. As a result, the number of mobile users

---
[*] lfuente@unizar.es

increased in European countries during that decade, reaching penetration rates close to 100% at the beginning of the 21st century. Of course, with these rates, subscriber growth started to become more moderate. In spite of these demographic restrictions to growth, the compatibility among the digital mobile systems of different countries allowed firms to exploit scale and scope economies by operating in several markets at the same time. In order to improve performance, European wireless operators started internationalizing their businesses all over the world, especially in Europe, where GSM was imposed by supranational institutions. The introduction of the UMTS standard in the first years of the 21st century gave an additional chance for the international diversification of European firms through the awarding of new licenses that allowed firms to enter into new markets. Consequently, several of the most important European mobile operators, such as Vodafone, T-Mobile, Orange, Movistar, Telia Sonera and Telenor, expanded their scope to other national markets with compatible digital standards, usually through the acquisition of the majority ownership of local mobile network operators (MNO, hereafter) that had previously obtained 2G and 3G licenses.

This chapter tries to offer a comprehensive description of the internationalization of these European mobile companies from 1998 to 2008 by differentiating between expansion across European OECD countries and expansion into the rest of the world.[1] One of our objectives is to observe the evolution of the internationalization of the main operators in order to identify differences among the most internationalized operators, by focusing on the number of subscribers, the number of countries in which they are present and, finally, the geographic scope of their expansion. Furthermore, we wish to study their entry modes and rebranding, two strategies related to internationalization.

The remainder of the chapter is organized as follows. Section 1 introduces the mobile communications industry in the European context with a description of the origins and evolution of the standardization process that encouraged the internationalization of European mobile operators. Section 2 describes the expansion of the main European wireless firms between 1998 and 2008 in the European OECD countries. Special attention is devoted to the number of countries in which each firm is present, the number of subscribers and the geographic scope of each operator in order to determine the main patterns of internationalization. The entry and rebranding strategies are also analyzed in this section, confirming the existence of differences among European operators in how they expand their boundaries. In Section 3, the criteria of number of countries and geographic scope are taken up again in order to describe the differences in the degree of internationalization of European operators in the rest of the world. Finally, the main conclusions derived from the previous sections are summarized.

---

[1] The European OECD countries have been considered as a special region for the study of the expansion of European mobile operators since: a) most internationalized operators in Europe have their origin in these countries; b) European institutions imposed the adoption of the GSM standard in most of these countries; and c) they have similar economic, politic and social characteristics. The European OECD countries that have been considered are Austria, Belgium, the Czech Republic, Denmark, Finland, France, Germany, Greece, Hungary, Ireland, Italy, the Netherlands, Norway, Poland, Portugal, Spain, Sweden, Switzerland and the United Kingdom. Iceland, Luxembourg and Slovakia have been excluded from the European OECD countries because of the lack of data by operator in these countries.

1. ORIGINS AND EVOLUTION OF EUROPEAN MOBILE COMMUNICATIONS

The mobile communications industry has attracted the attention of both scholars and practitioners. It has been selected as the research setting in the literature to illustrate topics such as network effects (Doganoglu and Gryzybowski, 2007; Grajek, 2010; Fuentelsaz, Maicas and Polo, 2010), technology diffusion (Gruber and Verboven, 2001; Jang, Dai and Sung, 2005) and users' choices (Birke and Swann, 2006; Maicas, Polo and Sese, 2010). This is not surprising given the social and economic importance of the sector in our society (Fuentelsaz, Maicas and Polo, 2008). The penetration rate of this technology reached 100% in Europe in 2005. If we observe Figure 1, we can see that the penetration of mobile technology in Europe has increased 500% in only 10 years. At the end of 1998, only around 25% of the population of the OECD European countries had a mobile handset, whereas 125% had adopted this technology in 2008. Nevertheless, some differences exist across countries. Figure 1 also shows the minimum and maximum penetration rates for each year, giving an idea of the variance across countries.

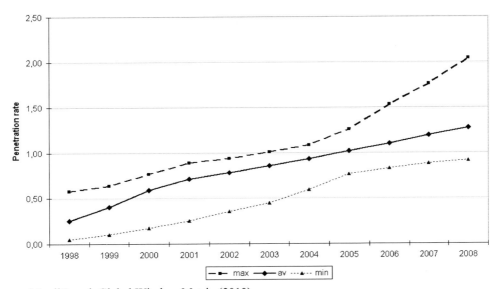

Source: Merril Lynch Global Wireless Matrix (2010).

Figure 1. Penetration Rates of Mobile Communications in European Countries (1998-2008).

The historical analysis of the wireless industry is highly connected to the technology evolution. The first mobile telephone system in Europe was commercialized by Swedish Telecom in 1956. Later, mobile systems were launched in Germany (1959), United Kingdom (1959) and other European countries in the 60's and 70's (Gruber, 2005). In spite of these first attempts, the industry was not really developed until the 80's with the introduction of analogue systems.

The analogue systems were based on radio waves that varied in frequency and technology across countries (Gruber, 2005). As can be seen in Figure 2, the early 80's show a substantial growth in the number of subscribers, probably due to the novelty of the technology. Nevertheless, the number of users was still moderate in this first stage in comparison with the following years. The academic literature has suggested several reasons for this low number of users, including the high prices in a monopoly regimen, the inexistence of a critical mass and the technology restrictions derived from incompatible standards between the networks of different countries. As an example of the latter, it can be mentioned that the independent development of mobile systems in each country made international roaming impossible in a European Union area that was moving towards fuller integration (Fuentelsaz et al., 2008).

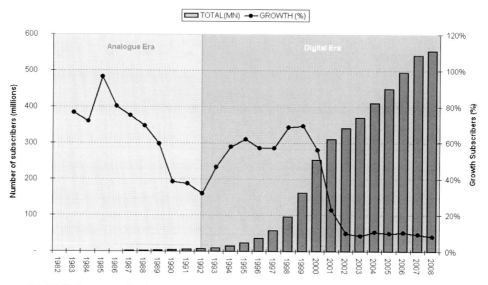

Source: OECD Telecommunications Database (1982-2005) and Merril Lynch Global Wireless Matrix (2006-2008).

Figure 2. Number of Mobile Subscribers in European Countries (1982-2008).

Consequently, there arose increasing concern about the necessity of making mobile systems compatible. As a result, the Group Special Mobile (GSM) was created in 1982 to work on the development of a compatible standard across European countries aimed at improving the quality and efficiency of phone services. Although the first agreement to implement this GSM standard was signed in September 1987 by 14 operators from 13 countries (Hillebrand, 2002), its commercial take-off occurred in 1992. This year can be considered as the beginning of the *digital era* of mobile technology in Europe. As can be seen in Figure 2, the number of subscribers started to grow radically from then on. The success of GSM was based on several advantages over the analogue system that Fuentelsaz et al. (2008) summarize as a more efficient use of the radio spectrum, cost advantages related to microelectronic technology, the possibility of international roaming, the exploitation of scale economies by manufacturers and a better distribution of the sunk costs of RandD among the European countries. The fast increase in the number of subscribers was accompanied by a growth in wireless technology penetration that, in the early 2000's, was close to 100%. Even after the introduction of UMTS in 2003, the annual growth in the number of subscribers and in penetration rate was maintained at around 10%, as can be seen in Figures 1 and 2.

The introduction of a digital standard was not only an improvement in the technology, but also encouraged the internationalization of wireless operators. Gerpott and Jakopin (2005:636) explain that "internationalization of telcos received an additional impetus with the licensing of digital mobile networks in numerous countries with most of these networks using the Global System for Mobile Communication (GSM) standard".[2] For the first time in wireless telecommunications in Europe, it was possible to exploit scale and scope economies more efficiently by operating in several countries at the same time. This allowed cost reductions, with the subsequent impact on performance. European operators started operating outside their domestic markets in order to continue growing. National boundaries were exceeded and, in the last years of the 20th century the internationalization of European firms was a reality. The introduction of UMTS gave a second boost to the internationalization of European operators by giving them the possibility of obtaining 3G licenses that allowed them to enter into new markets.

2. THE EXPANSION PROCESS OF EUROPEAN MOBILE OPERATORS IN EUROPEAN OECD COUNTRIES

This section offers a detailed description of the international expansion of the main European mobile firms in European OECD countries. First, we present a brief description of the expansion of the main operators that allows us to identify different patterns of internationalization depending on the timing, degree and geographic scope of the internationalization. Second, we describe two of the main expansion strategies followed by mobile firms in this context, namely, entry mode and rebranding decisions. Finally, once the most internationalized European operators have been identified, we analyze the evolution of their importance in the OECD countries by observing the evolution of their European market shares.

The Internationalized European Operators in European OECD Countries

We have selected the European firms that operated in two or more European OECD countries in 2008 and, after that, we calculate the number of countries in which each operator was present from 1998 to 2008. Table 1 provides information about the number of countries in which each firm was operating from 1998 to 2008. The information has been collected from the yearly reports of the European mobile operators. An ownership of over 50% of the national MNO by the international group was required for the group to be considered as present in a country. If intermediate firms exist, the international group has to own more than 50% of the intermediate firm(s), and the latter must own more than 50% of the MNO. With these limits, the existence of effective control is guaranteed.[3] International groups that have exclusively expanded outside the European area under analysis have not been analyzed. We

---

[2] GSM was initially the abbreviation of Group Special Mobile but it was changed to refer the standard, Global System for Mobile Communications, when the group was renamed Standard Mobile Group (SMG).
[3] In some cases, effective control has been supposed when the international group makes it explicit in the annual report.

can see that only a few firms are present in more than one country, namely, Vodafone, T-Mobile, Orange, Movistar, Telia Sonera, Telenor, KPN, TDC and Wind.

Table 1. Number of European Countries by Operator (1998-2008)*

	1998	1999	2000	2001	2002	2003	2004	2005	2006	2007	2008
Vodafone	3	5	9	10	10	10	10	11	10	10	10
T-Mobile	1	3	4	5	6	6	6	6	7	7	7
Orange	3	4	6	6	6	6	6	7	7	6	6
Movistar	1	1	1	1	1	1	1	2	5	5	5
Telia Sonera[4]	3	3	4	4	4	4	4	4	5	5	5
Telenor	1	1	2	2	3	3	3	3	4	4	4
KPN	1	1	2	3	3	3	3	3	3	3	3
TDC	1	1	1	2	2	2	2	2	2	2	2
Wind	-	1	1	1	1	1	1	1	1	2	2

* Figures as of December of each year (June in 2008).
Source: Own elaboration based on yearly corporate reports.

Table 2 shows the countries in which each firm was present in June 2008. Depending on the degree and scope of internationalization, we identify two groups. The first one (that will be called highly-internationalized operators) is made up of Vodafone, T-Mobile, Orange, Movistar, Telia Sonera and Telenor. All of them are present in at least four European countries and, as we will see in Section 3, they have also expanded their activities outside the European OECD countries. We will call the remaining firms (KPN, TDC and Wind) low-internationalized but they will not be considered in the remainder of our analysis.

Once we have identified our target firms, it is also interesting to note that the timing of their internationalization is completely different. While Vodafone, T-Mobile, Orange, Telia Sonera and Telenor started to expand in the early years of our observation window, Movistar waited until the appearance of the UMTS technology to widen its geographic scope.

Vodafone, which operated in the United Kingdom, the Netherlands and Greece before 1998, continued its internationalization by entering into Hungary, Portugal, Germany, Spain, Sweden and Ireland from 1999 to 2001. Later, the acquisition of a MNO in the Czech Republic in 2005 was counteracted by the selling of the Swedish operator in 2006 to Telenor.

T-Mobile started its internationalization in 1999 by acquiring operators in the United Kingdom and Austria. In the following three years, MNOs of Hungary, the Czech Republic and Netherlands were acquired by the group. The next entry took place in 2006 in Poland with the acquisition of Era. Finally, in 2007, T-Mobile Netherlands acquired Orange Netherlands from France Telecom, the two societies merging under the T-Mobile brand.

---

[4] The merger of Telia and Sonera took place in December 2002. For the previous period, the number of countries that appear in the Table 1 refers to sum of countries where either firm was present, although they were independent.

Table 2. Countries by Mobile Operator

International Group	Countries	Entry into the group (exit)
VODAFONE	United Kingdom	Before 1998
	Netherlands	Before 1998
	Greece	Before 1998
	Hungary	1999
	Portugal	1999
	Germany	2000
	Spain	2000
	Italy	2000
	Sweden	2000 (2006)
	Ireland	2001
	Czech Republic	2005
T-MOBILE	Germany	Before 1998
	United Kingdom	1999
	Austria	1999
	Hungary	2000
	Czech Republic	2001
	Netherlands	2002
	Poland	2006
ORANGE (FT)	France	Before 1998
	Belgium	Before 1998
	Denmark	Before 1998 (2004)
	Netherlands	1999 (2007)
	Switzerland	2000
	United Kingdom	2000
	Poland	2005
	Spain	2005
MOVISTAR	Spain	Before 1998
	Czech Republic	2005
	Germany	2006
	United Kingdom	2006
	Ireland	2006

Table 2. (Continued)

International Group	Countries	Entry into the group (exit)
TELIA SONERA	Sweden	Before 1998
	Finland	Before 1998
	Denmark	Before 1998
	Norway	2000
	Spain	2006
TELENOR	Norway	Before 1998
	Denmark	2000
	Hungary	2002
	Sweden	2006
KPN	Netherlands	Before 1998
	Germany	2000
	Belgium	2001
TDC	Denmark	Before 1998
	Switzerland	2001[5]
WIND	Italy	Before 1998
	Greece	2007

Source: Own elaboration based on yearly corporate reports.

Orange (or France Telecom) was present in France, Belgium and Denmark in 1998. In 1999 and 2000 it entered into the Netherlands, Switzerland and the United Kingdom. After the acquisition of the British operator, France Telecom started commercializing its mobile products under the name of Orange.[6] In 2005, Orange started operating in Poland and Spain, although these new entries took place approximately at the same time as they abandoned the Danish and Dutch markets in 2004 and 2007, respectively.

Telia Sonera and Telenor started internationalizing in the late 90's and at the beginning of the 21st century. A second stage of internationalization occurred in 2005-2006 in the UMTS era. However, compared to Vodafone, Orange and T-Mobile, their degree of internationalization is lower. *Telia Sonera* was founded in December 2002 following the

---

[5] Acquisition in December 2000 but with legal effect from January 2001.
[6] Orange.plc was founded by Hutchison Telecom in United Kingdom in 1994 and acquired by Mannesmann in 1999. In April 2000, Vodafone acquired Mannesmann. This acquisition gave it the control of two MNO's in United Kingdom, Vodafone and Orange. Vodafone sold Orange to France Telecom in August 2000. The French company previously commercialized their mobile services under the Itineris, OLA and Mobicarte brands but, from June 2001, it started to commercialize its mobile products taking advantage of the fame of the Orange brand.

merger between the Swedish Telia and the Finnish Sonera. Previously, Telia had been operating in Denmark and Norway. After the merger, Telia Sonera integrated Orange Denmark under the Telia Sonera brand in 2004 and acquired Xfera in Spain in 2006.

Telenor had been only present in Norway until 2000 when the firm started operating in Denmark. In 2002, the firm extended their scope to Hungary with the acquisition of Pannon. As we have said, in 2006, Telenor acquired Vodafone Sweden.

The European diversification of *Movistar* took place in the mid-2000's with the introduction of UMTS, by entering into the Czech Republic, Ireland, the United Kingdom and Germany during 2005 and 2006. So, Movistar took the decision to enter into European markets later than the operators described above.

Internationalization from a Number of Subscribers' Perspective

The number of countries is not the only variable to take into account when the internationalization process is analyzed. Given that the size and economic importance of these countries is different, we also consider the total number of subscribers to each firm. Figure 3 shows the evolution in the number of subscribers across Europe of the highly-internationalized operators. Clearly, Telia Sonera and Telenor present a lower degree of internationalization over the 10-year period, with no more than 20 million users in 2008, while the number of subscribers to the remaining firms ranges from 60 to 120 million. This difference is mainly due to two reasons. First, the larger operators are present in a greater number of countries. Second, they operate in one or more of the most populated countries in Europe. In any case, the evolution in the number of subscribers is similar to the evolution in the number of countries.

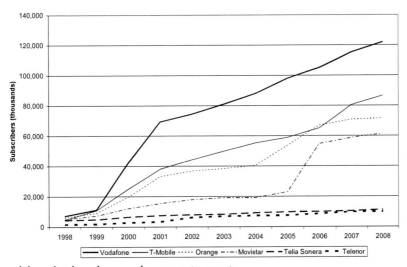

Source: Own elaboration based on yearly corporate reports.

Figure 3. number of Subscribers in Europe (1998-2008).

Figures 4, 5 and 6 show the three patterns of internationalization in terms of the degree and timing of internationalization. These figures compare the growth in the number of subscribers by year and operator and compare it with the average growth of European mobile users.

Vodafone, T-Mobile and Orange (Figure 4) made the largest expansion investments at the end of the 20th and beginning of the 21st centuries. Although several entries took place in 2005 and 2006, the growth in the number of users by operator in the UMTS era was close to the European average, especially in the case of Vodafone.

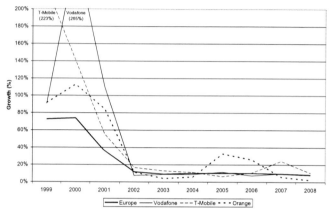

Source: Own elaboration based on yearly corporate reports.

Figure 4. Vodafone, T-Mobile, Orange and Europe: Average Subscribers Growth (1998-2008).

The most important investments in the international expansion of Telia Sonera and Telenor (Figure 5) took place during the first part of our observation window with the entry of Telia Sonera into Norway (2000) and the entry of Telenor into Denmark (2000) and Hungary (2002). The latest entries in Spain (2006) and Sweden (2006), by Telia Sonera and Telenor, respectively, had a lower impact on the number of subscribers in their international installed base.

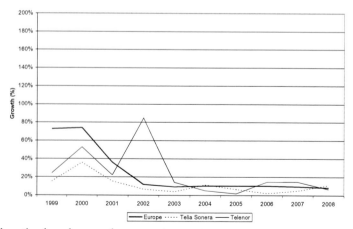

Source: Own elaboration based on yearly corporate reports

Figure 5. Telia Sonera, Telenor and Europe: Average Subscribers Growth (1998-2008).

Finally, the internationalization of Movistar (Figure 6) started after the introduction of UMTS in 2003, by entering into the Czech Republic (2005), Germany, the United Kingdom and Ireland (2006). These entries allowed it to become one of the largest operators in Europe in terms of number of subscribers in only five years. Before these entries, the growth in the number of users of Movistar was close to the European average.

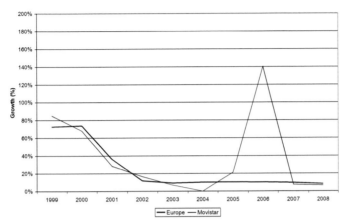

Source: Own elaboration based on yearly corporate reports

Figure 6. Movistar and Europe: Average Subscribers Growth (1998-2008).

It is important to note that, in all cases, the growth in the number of subscribers of each operator tends to be very similar to the growth in the number of subscribers in Europe – except in the years in which a new MNO is acquired. This shows that the growth in the number of subscribers in Europe has corresponded with the expansion of the highly-internationalized operators.

Table 3 synthesizes the different patterns of internationalization followed by the main European firms considering the two dimensions that we have considered in our analysis: degree and timing of internationalization. Table 3 identifies three groups. Vodafone, T-Mobile and Orange are characterized by an early and high degree of internationalization. Telia Sonera and Telenor have an early and moderate degree of internationalization. Finally, Movistar is characterized by a late and high degree of internationalization.

Table 3. Classification of Highly-internationalized European Operators by Timing of Entry and Degree of Internationalization in OECD Europe

		Timing of internationalization	
		Early ← - + → Late	
Degree of Internationalization	High ↑ + - ↓ Low	Vodafone T-Mobile Orange Telia Sonera Telenor	Movistar

Source: Personal compilation based on annual corporative reports

Internationalization from a Geographic Perspective

Finally, we complete the description of the expansion of European firms across Europe by analyzing the countries in which they chose to invest from a geographic perspective. Figures 7-12 show maps of the countries in which each operator is present. The colours show the timing of entry into each country, differentiating between entries before our period of study (before 1998), entries during the GSM era (1999-2003) and entries during the UMTS era (2004-2008).

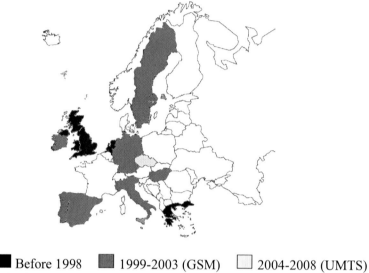

■ Before 1998   ▨ 1999-2003 (GSM)   ☐ 2004-2008 (UMTS)

Source: Own elaboration based on yearly corporate reports.

Figure 7. Vodafone (1998-2008).

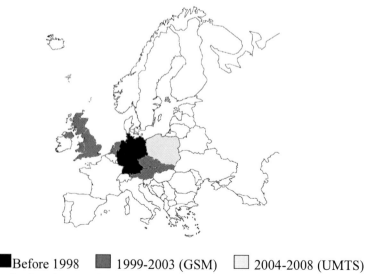

■ Before 1998   ▨ 1999-2003 (GSM)   ☐ 2004-2008 (UMTS)

Source: Own elaboration based on yearly corporate reports.

Figure 8. T-Mobile (1998-2008).

■ Before 1998   ■ 1999-2003 (GSM)   □ 2004-2008 (UMTS)

Source: Own elaboration based on yearly corporate reports.

Figure 9. Orange (1998-2008).

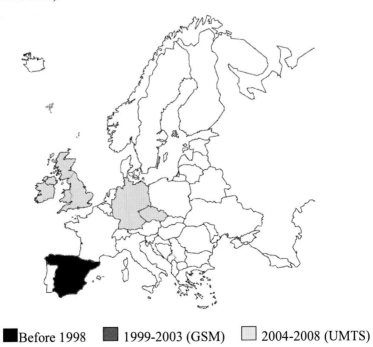

■ Before 1998   ■ 1999-2003 (GSM)   □ 2004-2008 (UMTS)

Source: Own elaboration based on yearly corporate reports.

Figure 10. Movistar (1998-2008).

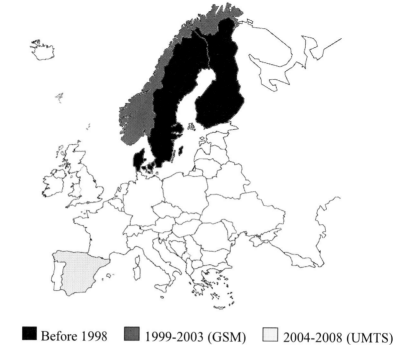

■ Before 1998   ■ 1999-2003 (GSM)   □ 2004-2008 (UMTS)

Source: Own elaboration based on yearly corporate reports.

Figure 11. Telia Sonera (1998-2008).

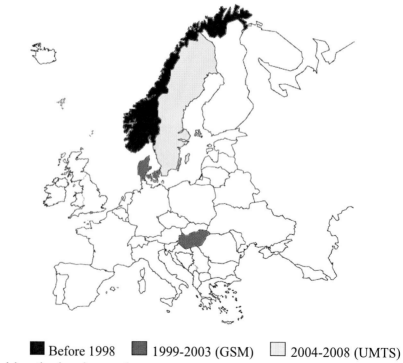

■ Before 1998   ■ 1999-2003 (GSM)   □ 2004-2008 (UMTS)

Source: Own elaboration based on yearly corporate reports.

Figure 12. Telenor (1998-2008).

As can be appreciated, there are two clear tendencies in the geographic expansion. Vodafone, T-Mobile, Orange and Movistar have paid special attention to West, Central and South Europe. The only market where all them are present is the United Kingdom. But three of them coincide in Spain, Germany, the Czech Republic and the Netherlands. On the other hand, Telia Sonera and Telenor have focused especially on North Europe (Finland, Sweden, Norway, and Denmark). This distribution of markets is confirmed by the acquisition of Vodafone Sweden by Telenor (2006) and of Orange Denmark by Telia Sonera (2004). With these operations, the Northern operators achieved control of the Nordic markets.

Another appreciation can be found when we observe the countries chosen by the firms to expand their boundaries. With the sole exception of Movistar, the other European operators have internationalized by entering countries close to their domestic market. For example, Vodafone (from the United Kingdom) operates in the Netherlands (before 1998) and Ireland (2001); T-Mobile (from Germany) has entered Austria (1999), the Czech Republic (2001), the Netherlands (2002) and Poland (2006); Orange (from France) operates in Belgium (before 1998), the United Kingdom (2000), Switzerland (2002) and Spain (2005); Telia Sonera (from Finland and Sweden) has entered Denmark (before 1998) and Norway (2000); and Telenor (from Norway) has expanded to Denmark (2000) and Sweden (2006).

Strategic Decisions in the European Internationalization Process: Entry Mode

Together with the internationalization process, the evidence also shows an evolution in the entry mode during the period we are considering. At the beginning, firms usually entered into foreign markets through greenfield investments (as in the case of Vodafone in Hungary or the joint-venture between KPN and Orange in Belgium with KPN Orange, both in 1999) or by acquiring a minority share of an established operator. For example, in 1998, T-Mobile was present in Germany with a majority ownership of the MNO and in 6 other countries with a minority ownership. However, the landscape in 2008 is completely different. Minority investments have disappeared. Gerpott and Jakopin (2005:648) observe that "based on the minority investment experience until the mid-1990s some of the European MNO began to implement new majority takeovers of foreign firms, which had already been in the MNO business, and to transform several of their minority stakes into majority holdings". These authors conclude that the efficient integration of international business was impossible without a majority ownership.

Table 4 confirms this tendency towards majority ownership. It compares the number of countries, in 1998 and 2008, where each operator had total (above 50%) or partial control of each foreign subsidiary. In 2008, almost 90% of the firms had total control of them (107 out of 120) whereas, ten years earlier, this figure was only 38%. Table 4 also offers a detailed picture of each firm, with similar results to the ones commented.

Having confirmed that majority investment is the main entry mode, we focus on this entry behaviour. Table 5 shows the moment at which each European operator obtains the majority ownership of the acquired national MNOs between 1998 and 2008. Additionally, we show the percentage of ownership immediately following the acquisition, the greenfield investment or merger.

Table 4. Number of Countries by Type of Ownership (1998-2008)

	Majority Ownership			Minority Ownership		
	1998	2008	Change	1998	2008	Change
Vodafone	6	17	↑	2	4	↑
T-Mobile	1	12	↑	6	0	↓
Movistar	3	18	↑	6	2	↓
Telenor	1	11	↑	9	2	↓
Orange	11	28	↑	10	2	↓
TDC	2	2	=	6	0	↓
KPN	1	3	↑	0	0	=
Telia Sonera	3	14	↑	8	3	↓
Wind	1	2	↑	0	0	=
TOTAL	29	107	↑	47	13	↓

Source: Personal compilation based on annual corporative reports

Table 5. European Operators: Acquiring Effective Control

Country	Entry into the group 1998-2008 (exit)	Entry mode
VODAFONE	Origin: United Kingdom	
Hungary	1999	Greenfield (50.1%)
Portugal	1999	Acquisition (50.9%)
Germany	2000	Acquisition (98.6%)
Spain	2000	Acquisition (73.8%)
Italy	2000	Acquisition (76.1%)
Sweden	2000 (2006)	Acquisition (71.1%)
Ireland	2001	Acquisition (100%)
Czech Republic	2005	Acquisition (100%)
T-MOBILE	Origin: Germany	
United Kingdom	1999	Acquisition (100%)
Austria	1999	Acquisition (100%)
Hungary	2000	Acquisition (79.4%)
Czech Republic	2001	Acquisition (56.0%)
Netherlands	2002	Acquisition (100%)
Netherlands	2007	Merger[7]
Poland	2006	Acquisition (97.0%)

---

[7] In October 2007, T-Mobile Netherlands acquired Orange Netherlands from France Telecom. The mobile segment was integrated into T-Mobile Netherlands.

Country	Entry into the group 1998-2008 (exit)	Entry mode
ORANGE (FT)	Origin: France	
Netherlands	1999 (2007)	Greenfield (80.0%)
Switzerland	2000	Greenfield (87.5%)
United Kingdom	2000	Acquisition (100%)
Poland	2005	Acquisition (100%)[8]
Spain	2005	Acquisition (80.0%)
MOVISTAR	Origin: Spain	
Czech Republic	2005	Acquisition (69.4%)
Germany	2006	Acquisition (100%)
United Kingdom	2006	Acquisition (100%)
Ireland	2006	Acquisition (100%)
TELIA SONERA	Origin: Sweden- Finland	
Norway	2000	Acquisition (100%)
Finland	2002	Merger[9]
Denmark	2004	Merger[10]
Spain	2006	Acquisition (76.6%)
TELENOR	Origin: Norway	
Denmark	2000	Acquisition (53.5%)
Hungary	2002	Acquisition (100%)
Sweden	2006	Acquisition (100%)[11]
KPN	Origin: Netherlands	
Germany	2000	Acquisition (77.5%)
Belgium	2001	Acquisition (100%)[12]
Netherlands	2005	Merger[13]

---

[8] Until 2003, France Telecom held 34% of the TP Group, owner of 100% of PTK (Orange Poland). In 2003, France Telecom increased its ownership to 47.5% of the TP Group. As the change of brand took place in 2005, we have assumed that, from then on, FT has had an effective control over the TP Group and, consequently, over PTK. The annual reports of 2007 and 2008 refer to this majority control.

[9] The international merger of Telia and Sonera implied the disappearance of Telia as an independent firm in Finland in 2002.

[10] Acquisition of 100% of Orange Denmark by Telia Sonera in 2004.

[11] Acquisition of 100% of Vodafone Sweden by Telenor in January 2006.

[12] KPN Orange was founded in 1999 as a joint-venture of the British Orange (50.0%) and the Dutch KPN (100%). But, after the acquisition of Orange by France Telecom, KPN Orange was totally acquired by KPN in 2001, with the subsequent change in the commercial brand of the mobile network to Base in 2002. Thus, KPN Orange was initially a greenfield although KPN did not have total control of the firm, which explains why we consider 2001 to be the entry date of the Belgium MNO into the international group.

[13] Acquisition of 100% of Telfort in the Netherlands by KPN in October 2005. This firm is integrated into KPN Netherlands.

Table 5. (Continued)

Country	Entry into the group 1998-2008 (exit)	Entry mode
TDC	Origin: Denmark	
Switzerland	2001	Acquisition (78.1%)
WIND	Origin: Italy	
Greece	2007	Acquisition (100%)

Source: Own elaboration based on yearly corporate reports.

We can observe that, in most cases, the international wireless groups take control of national MNOs with a percentage near to 100%. Several exceptions can be found with the acquisition of Vodafone Hungary and Vodafone Portugal (50.1% and 50.9%, respectively), T-Mobile Czech Republic (56.0%) and Telenor Denmark (53.5%).

Although the operators have tended to maintain their participation in the acquired firms, several changes have taken place in the 10-year period that we consider. The first one results in a reconfiguration of the market in the Nordic countries and starts with the merger between Telia (Sweden) and Sonera (Finland) in 2002. In 2004, the resulting company acquired Orange Denmark, merging under the Telia brand In January 2006, Vodafone Sweden was acquired by Telenor, which allowed this company to enter into the Swedish market. These acquisitions gave control of the North European market to the Northern operators.

The second change began in October 2005 with the merger of Telfort and KPN Netherlands (with the subsequent disappearance of the former which was integrated into KPN Netherlands) and continued in October 2007 when France Telecom sold Orange Netherlands to T-Mobile, which integrated it into its business. These mergers changed the market structure in the Netherlands where the five companies initially operating (KPN, Vodafone, Orange, Telfort and T-Mobile) were reduced to three (KPN, Vodafone and T-Mobile) in only two years.

Strategic Decisions in the European Internationalization Process: Rebranding

Different patterns can be observed in the rebranding decisions of the international wireless firms in Europe. First, the biggest international operators choose to create *global brands* as one of the cornerstones of their internationalization strategies. Global brands are defined as "brands that consumers can find under the same name in multiple countries with generally similar and centrally coordinated marketing strategies" (Steenkamp, Batra and Alden, 2003: 53). These authors highlight the main reasons that firms develop global brands: they allow the firms to exploit scale and scope economies in RandD, manufacturing and marketing and consumers prefer brands with a "global image" to local competitors, even when their quality and value are not superior. These authors show that global brands are positively related to perceived prestige and brand quality.

Table 6 shows that, at the beginning of the 21st century, mobile operators such as Vodafone, Orange and T-Mobile were the first to make efforts to extend their global brand across countries. For example, Vodafone started to expand its brand name to the mobile operators that it had acquired in Sweden (2000), Portugal and Spain (2001) and the Netherlands, Greece, Germany and Ireland (2002). In the case of T-Mobile, in 2002, the firm changed its name from "T-Mobil" to "T-Mobile" in order to make it look English. Its rebranding process started with the rebranding of the operators from Germany, the United Kingdom, Austria and the Czech Republic (2002), the Netherlands (2003) and Hungary (2004). Another example is Orange, the brand of France Telecom. This case is paradigmatic because the French group decided to extend the name of the British operator that it acquired in 2000 as its global brand, starting the rebranding process in 2001 in France and Denmark. Thus, the international groups have tended to use an English name to become internationally known. In the case of France Telecom, the selection of "Orange" was especially strategic because it is an English word (suitable for international expansion) but also a French word (which would not cause rejection from the French-speaking customers of France Telecom).

Table 6. Rebranding of European Operators (1998-2008)*

Country	Entry into the group (exit)	Previous Brand	Current Brand	Brand Change
VODAFONE				
United Kingdom	Before 1998	-	Vodafone	-
Netherlands	Before 1998	Libertel	Vodafone	2002
Greece	Before 1998	Panafon	Vodafone	2002
Hungary	1999	-	Vodafone	-
Portugal	1999	Telecel	Vodafone	2001
Germany	2000	D2-Netz	Vodafone	2002
Spain	2000	Airtel	Vodafone	2001
Italy	2000	Omnitel	Vodafone	2003
Sweden	2000 (2006)	Europolitan	(Vodafone)	2000
Ireland	2001	Eircell	Vodafone	2002
Czech Republic	2005	Oskar	Vodafone	2006
T-MOBILE				
Germany	Before 1998	T-Mobil	T-Mobile	2002
United Kingdom	1999	One2One	T-Mobile	2002
Austria	1999	max.mobil	T-Mobile	2002
Hungary	2000	Westel	T-Mobile	2004
Czech Republic	2001	Paegas	T-Mobile	2002
Netherlands	2002	Ben	T-Mobile	2003
Poland	2006	-	Era	-

Table 6. (Continued)*

Country	Entry into the group (exit)	Previous Brand	Current Brand	Brand Change
ORANGE (FT)				
France	Before 1998	Itineris OLA, Mobicarte	Orange	2001
Belgium	Before 1998	-	Mobistar	-
Denmark	Before 1998 (2004)	Mobilix	Orange	2001
Netherlands	1999 (2007)	Dutchtone	Orange	2003
Switzerland	2000	-	Orange	-
United Kingdom	2000	-	Orange	-
Poland	2005	Idea	Orange	2005
Spain	2005	Amena	Orange	2006
MOVISTAR				
Spain	Before 1998	-	Movistar	-
Czech Republic	2005	Eurotel	O2	2006
Germany	2006	-	O2	-
United Kingdom	2006	-	O2	-
Ireland	2006	-	O2	-
TELIA SONERA				
Sweden	Before 1998	-	Telia	-
Finland	Before 1998	-	Sonera	-
Denmark	Before 1998	-	Telia	-
Norway	2000	-	Netcom	-
Spain	2006	-	Yoigo	-
TELENOR				
Norway	Before 1998	-	Telenor	-
Denmark	2000	-	Sonofon[14]	-
Hungary	2002	-	Pannon[15]	-
Sweden	2006	(Vodafone)	Telenor	2006
KPN				
Netherlands	Before 1998	-	KPN	-
Germany	2000	-	E-plus	-
Belgium	2001	KPN Orange	Base	2002

[14] Sonofon changed its name to Telenor in June 2009.
[15] Pannon changed its name to Telenor in May 2010.

Country	Entry into the group (exit)	Previous Brand	Current Brand	Brand Change
TDC				
Denmark	Before 1998	-	TDC	-
Switzerland	2001	-	Sunrise	-
WIND				
Italy	Before 1998	-	Wind	-
Greece	2007	TIM	Wind	2007

Source: Own elaboration based on yearly corporate reports.
* Brand in brackets means that the brand had disappeared in that country in 2008.

Another firm that has lately been trying to strengthen its global brand is Telenor, which rebranded Vodafone Sweden as Telenor (2006). Although outside the period of study (1998-2008), Telenor has decided to rebrand the mobile operators of Denmark (2009) and Hungary (2010) with its international brand. Wind also operates in Italy and Greece under the Wind brand, although its importance in terms of international presence is much lower.

The case of Movistar (Spain) is peculiar. This firm has two important brands which have been extended in terms of the geographical market. Movistar is the brand that operates in Spain and in South American countries. However, in its expansion across Europe, the firm has selected O_2 as its international brand. In this case, the strategy has consisted in not changing the brand of the operators of United Kingdom, Germany and Ireland acquired from the mmO2 group in 2006 since it was already a consolidated brand internationally speaking. However, the O2 brand was also given to the mobile operator from the Czech Republic (previously, Eurotel). This is a sign of the interest of Movistar in extending O2 as the European brand of the Spanish group.

It is important to note that giving a global dimension to the brands is becoming a more usual strategy, since it can positively influence firm image and reputation. Global brand strategy has accompanied the majority stakeholder acquisitions to improve reputation and brand value, especially in the highly-internationalized firms such as Vodafone, T-Mobile, Orange and Movistar(O2).

The other European operators under study have hardly made any effort to extend a single brand across European countries. For example, Telia Sonera also operates with the brands Netcom (Norway) and Yoigo (Spain). KPN operates under the KPN brand (Netherlands), E-Plus (Germany) and Base (Belgium). KPN had operated under the KPN Orange brand in Belgium until 2002 since it was a joint-venture between KPN and Orange. But, with the acquisition of the whole firm by KPN in 2001, it was decided to change the brand to eliminate the references to Orange and to KPN by establishing a completely different brand, namely, Base. All this shows that KPN has little interest in creating a global brand. TDC operates in Denmark with its original brand TDC but, in Switzerland, under the Sunrise brand. So, for these medium and low internationalized groups, the global brand is not considered to be a valuable intangible that must be promoted.

The European Markets Shares of European Operators

In the previous sections, we have analyzed the growth of the main firms that operate in the mobile communications industry in Europe. However, we have not provided information about their importance in the European context. This section goes a step further in our analysis and compares the market share of the highly-internationalized mobile companies in the first and last years of our observation window. Figure 13 confirms that the market share of the main firms has substantially increased in this ten-year period. In 1998, they accounted for 32% of the European market (measured by the number of subscribers). Vodafone (7.93%) and Orange (6.85%) led the market. Movistar (5.47%), T-Mobile (5.14%) and Telia Sonera (4.84%) had market shares of around 5% while Telenor's share is much lower (1.76%) (see graph A in Figure 13).

In 2008, the picture is completely different (see graph B in Figure 13). There are four companies that clearly lead the market (with a joint market share of over 61%). Vodafone is still the biggest firm and almost three times bigger than in 1998 (22.10%). T-Mobile (15.68%), Orange (12.91%) and Movistar (11.16%) have market shares of above 10%. Surprisingly, the relative size of Telia Sonera (2.05%) is now lower than in 1998, while the size of Telenor (1.79%) has hardly changed. The total market size of the highly-internationalized operators has more than doubled and accounts for almost 66% of the European market.

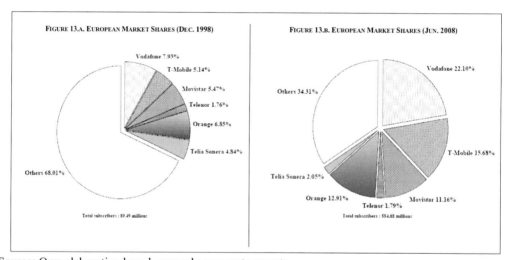

Source: Own elaboration based on yearly corporate reports

Figure 13. European Market Share by Operator.

INTERNATIONALIZATION AROUND THE WORLD: EUROPEAN OPERATORS BECOME GLOBAL PLAYERS

Having analyzed the internationalization of European firms across European OECD countries, we will now try to go a step further and provide a preliminary approach to the globalization of these firms in the rest of the world. As can be seen in Table 7, the European operators have greatly increased their global presence from 1998 to 2008. Orange is the European firm with the highest international presence: it offers its services in 28 countries (11 in 1998). Movistar and Vodafone (which are operating in 18 and 17 countries, respectively) have substantially increased their international scope (both firms had hardly began to internationalize before 1998). The other highly-internationalized operators were not operating in non-European OECD countries until 1999 (T-Mobile) and 2000 (Telia Sonera and Telenor). Some of the international European firms that we have analyzed in the previous sections (KPN, TDC and Wind) have chosen not to expand into the rest of the world. This is, indeed, one of the reasons why we have considered these firms to be low-internationalized operators in Section 2.

Table 7. Number of Countries by Operator (1998-2008)*

	1998	1999	2000	2001	2002	2003	2004	2005	2006	2007	2008
Orange	11(8)	14(10)	18(12)	20(14)	20(14)	20(14)	19(13)	22(15)	23(16)	24(18)	28(22)
Movistar	3 (2)	2 (1)	4 (3)	5 (4)	6 (5)	6 (5)	13(12)	14(12)	17(12)	18(13)	18(13)
Vodafone	6 (3)	8 (3)	13 (4)	15 (5)	15 (5)	15 (5)	15 (5)	17 (6)	16 (6)	17 (7)	17 (7)
Telia Sonera	3 (0)	3 (0)	4 (0)	4 (0)	10 (6)	10 (6)	11 (7)	11 (7)	12 (7)	14 (9)	14 (9)
T-Mobile	1 (0)	3 (0)	5 (1)	9 (4)	10 (4)	10 (4)	10 (4)	11 (5)	12 (5)	12 (5)	12 (5)
Telenor	1 (0)	1 (0)	2 (0)	3 (1)	5 (2)	6 (3)	7 (4)	9 (6)	10 (6)	10 (6)	11 (7)
KPN	1 (0)	1 (0)	2 (0)	3 (0)	3 (0)	3 (0)	3 (0)	3 (0)	3 (0)	3 (0)	3 (0)
TDC	2 (0)	2 (0)	2 (0)	3 (0)	3 (0)	3 (0)	3 (0)	4 (0)	4 (0)	2 (0)	2 (0)
WIND	-	1 (0)	1 (0)	1 (0)	1 (0)	1 (0)	1 (0)	1 (0)	1 (0)	2 (0)	2 (0)

*Dates referred to December (except 2008, June).
(Number of non European OECD countries in parenthesis).
Source: Own elaboration based on yearly corporate reports.

If we analyze the growth pattern of the firms that have a global presence in more detail (Table 8 details the countries where each firm is now active), we can observe that Orange currently offers its services in several countries in Africa, Latin America and Asia (Jordan and the Lebanon). The global expansion of Orange started even before 1998 (with a presence in Botswana, the Caribbean, the Ivory Coast and the Lebanon) and continued in the first years of our observation window: it entered Cameroon and Senegal in 1999, the Dominican Republic in 2000, Egypt and Reunion in 2001, Mali in 2003, Equatorial Guinea in 2005, Kenya in 2006 and the Central African Republic, Guinea Bissau, Niger and Vanuatu in 2008. Thus, Orange is clearly an early globalizer.

Movistar and Vodafone started their globalization a little bit later than Orange. It is true that they had some minor outside presence before 1998, but their global expansion really started at the beginning of the 21st century. Movistar started operating in Latin America in 2000, with its entry into Argentina and Peru (2000). It continued in 2001 (México), 2002 (El Salvador and Guatemala) and 2004 (Chile, Colombia, Ecuador, Nicaragua, Panamá, Uruguay and Venezuela).

Table 8. Non European OECD Countries by Operator (Jun. 2008)

International Group	Countries	International Group	Countries
ORANGE (FT) (France)	Botswana	MOVISTAR (Spain)	Argentina
	Cameroon		Chile
	Caribbean		Colombia
	Central African Republic		Ecuador
	Dominican Republic		El Salvador
	Egypt		Guatemala
	Equatorial Guinea		México
	Guinea		Nicaragua
	Ivory Coast		Panamá
	Jordan		Peru
	Kenya		Slovakia
	Lebanon		Uruguay
	Luxembourg		Venezuela
	Madagascar		
	Mali	VODAFONE (United Kingdom)	Australia
	Moldavia		Egypt
	Niger		India
	Reunion		Malta
	Romania		New Zealand
	Senegal		Romania
	Slovakia		Turkey
	Vanuatu		
T-MOBILE (Germany)	Croatia	TELIA SONERA (Sweden and Finland)	Azerbaijan
	Macedonia		Estonia
	Montenegro		Georgia
	Slovakia		Kazakhstan
	USA		Latvia
			Lithuania
TELENOR (Norway)	Bangladesh		Moldova
	Malaysia		Tajikistan
	Montenegro		Uzbekistan
	Pakistan		
	Serbia		
	Thailand		
	Ukraine		

Source: Own elaboration based on yearly corporate reports
(Country of origin in parenthesis)

Its expansion across European countries was reinforced by entry into Slovakia in 2007 under the O2 brand. With this last entry, Movistar is now present in 18 national markets, 13 of them in non-European OECD countries. Similarly, Vodafone was presence in 1998 in Australia, New Zealand and Malta, and expanded into Egypt (2000), Romania (2005), Turkey (2006) and India (2007). Thus, these operators can be considered to be recent globalizers.

Finally, there is another group of mobile firms that started their internationalization in non-European OECD countries in a later period. Telia Sonera, T-Mobile and Telenor belong to this group. Telia Sonera started internationalizing in 2002 with its entry into Azerbaijan, Georgia, Kazakhstan, Latvia, Lithuania and Moldova. It also entered into Estonia (2004) and Tajikistan and Uzbekistan (2007). With these entries, Telia Sonera has clearly expanded the boundaries of its business from North Europe to East Europe. T-Mobile expanded its boundaries to Slovakia in 2000. After entries into Macedonia, Croatia and the United States in 2001, its last entry took place in 2005 into Montenegro. Finally, Telenor has expanded its boundaries to Malaysia (2001), Ukraine (2002), Bangladesh (2003), Montenegro (2004), Pakistan (2005) and Serbia and Thailand (2006).

It is not easy to identify a clear pattern in the expansion of the different firms. However, cultural and historic reasons are often drivers of international expansion. Firms have tended to enter into countries with which they had maintained historical relationships, especially in the colonization stage. Orange (France) has expanded to African countries (e.g. Cameroon and Ivory Coast), Movistar (Spain) to Latin American (e.g. Argentina, Venezuela and Colombia) and Vodafone (United Kingdom) to previous British colonies around the world (Australia, Malta and India). It is interesting to note that European operators tend not to coincide in the same markets outside European OECD countries. This behaviour is quite different to that observed in their expansion across Europe, where the same firms tend to operate in the same markets.

A Map of the Internationalization Degree of the European Mobile Firms

In order to have a clearer picture of the internationalization processes of European operators inside and outside of the European OECD countries, Table 9 summarizes the information provided in the previous sections. It shows the positioning of each European operator in relation to two of the key variables that we have considered in this research: the timing and degree of internationalization. The main results that derive from the observation of Table 9 are the following:

a) Orange is an early-globalized operator.
b) Vodafone and Movistar started their globalization a little bit later than Orange. Vodafone started its European growth much earlier than Movistar.
c) T-Mobile, Telia Sonera and Telenor can be considered to be global late-internationalized operators but European early-internationalized operators. However, it could be useful to distinguish between T-Mobile and the two Northern operators since the latter have a lower presence in Europe.
d) KPN, TDC and Wind have been operating in two or three European OECD countries, but not in the rest of the world. Thus, they should be considered to be European low-internationalized operators.

Table 9. European Operators by degree of internationalization

			Global Internationalized			Only Europe
			Early	Recently	Late	
European Internationalized	High	Early	Orange	Vodafone	T-Mobile Telia Sonera Telenor	
		Late		Movistar		
	Low					KPN TDC Wind

Source: Own elaboration based on yearly corporate reports

CONCLUSIONS

This chapter has examined the internationalization process of the main European operators from 1998 to 2008. It has been shown that the introduction of digital standards encouraged the internationalization of these operators because it allowed them to exploit the scale and scope economies derived from the compatibility between the wireless networks of different countries.

The importance of international mobile groups has grown spectacularly in European OECD countries in the 10-year period under study, making international operators the most important agents of this sector in this area. However, the timing and degree of internationalization have not been the same for all of them. Offering an analysis by operator, this chapter shows that Vodafone, Orange, T-Mobile have been confirmed as early international operators with a higher degree of international presence in the European area. Telia Sonera and Telenor have also shown an early internationalization but have achieved a lower international installed base than the previous firms. In contrast, Movistar is a European late internationalized operator since its expansion across Europe took place after the introduction of the UMTS. In all cases, operators have tended to expand close to their domestic markets, which has led to many operators coinciding in the same countries. However, a division of areas has been observed: the Northern operators (Telia Sonera and Telenor) have achieved control of the North European countries, whereas Vodafone, Orange, T-Mobile and Movistar operate in West, Central and South Europe.

All these European operators have started an expansion into the rest of world whose timing is different to that observed in the European area. Orange was an early starter in its international expansion outside Europe (before 1998), entering, above all, ex-French colonies in Africa. Vodafone and Movistar had hardly started their internationalization before 1998 (at least, to a lesser degree than Orange). Yet again, history determined the countries chosen by these operators: Vodafone has expanded into South Asia and Oceania and Movistar has concentrated its efforts in South and Central American countries. T-Mobile and the Northern operators started operating outside OECD Europe after 1998, later than the previous firms.

They are located, above all, in East Europe. Thus, in non OECD European countries, European operators tend not to coincide in the same market.

This chapter has also attempted to describe two important strategies related to the internationalization of firms in European OECD countries. First, the entry mode of these operators has been analyzed. It has been observed that, although several greenfields have taken place, the European operators have tended to invest in existing MNOs. The evolution from minority to majority ownership of these firms took place during the 10-year period under study. International acquisitions have also occurred between these operators, such as the sale of Vodafone Sweden to Telenor and Orange Netherlands to T-Mobile. Thus, the dynamism in the corporative structures of telcos in Europe has been demonstrated.

The second strategy that has been described is rebranding. There have been two attitudes to the rebranding of MNOs that have been incorporated into the international corporations. On the one hand, some operators, such as KPN, TDC and Telia Sonera, have not achieved a sufficient degree of internationalization and prefer not to invest in creating a global brand. On the other hand, the highly-internationalized operators (such as Vodafone, Orange, T-Mobile, Movistar (O_2) and Telenor) find a global brand to be a valuable asset to compete internationally.

REFERENCES

Birke, D. and Swann, P. (2006). *"Network Effects and the Choice of Mobile Operator"*, Journal of Evolutionary Economics, 16 (1/2), 65 – 84

Curwen, P. and Whalley, J. (2008). *"The Internationalisation of Mobile Telecommunications. Strategic Challenges in a Global Market"*. Edward Elgar, Northampton, United States

Doganoglu, T. and Gryzbowski, L. (2007). *"Estimating Network Effects in the Mobile Telephony in Germany"*. Information Economics and Policy, 19 (1), 65 – 79

Fuentelsaz, L., Maicas, J.P. and Polo, Y. (2008). *"The evolution of mobile communications in Europe: the transition from the second to the third generation"*. Telecommunications Policy, 32 (6), 436-449

Fuentelsaz, L., Maicas, J.P. and Polo, Y. (2010). *"Switching costs, network effects and competition in the European mobile telecommunications industry"*. Information Systems Research, in press

Grajek, M. (2010). *"Estimating Network Effects and Compatibility: Evidence from the Polish mobile market"*. Information Economics and Policy, 22 (2), 130-143

Gerpott, T.J. and Jakopin, N.M. (2005). *"The degree of internationalization and the financial performance of European mobile network operators"*. Telecommunications Policy, 29 (8), 635-661

Gruber, H. and Verboven, F. (2001). *"The Diffusion of Mobile Telecommunications Services in the European Union"*. European Economic Review, 45 (3), 577 – 588

Gruber, H. (2005). *The Economics of Mobile Telecommunications*. Cambridge University Press, Cambridge, United Kingdom

Hillegrand, F. (2002). *GSM and UMTS. The Creation of Global Mobile Communications*. Willey. Sussex. England

Hitt, M.A., Hoskisson, R.E. and Kim, H. (1997). *"International Diversification: Effects on Innovation and Firm Performance in Product-Diversified Firms"*. Academy of Management Journal, 40(4), 767-798

Jang, S-L., Dai, S-C. and Sung, S. (2005). *"The Pattern and Externality Effect of Diffusion of Mobile Telecommunications: the Case of the OECD and Taiwan"*. Information Economics and Policy, 17 (2), 133-148

Jakopin, N.M. (2008). *"Internationalisation in Telecommunications Services Industry: Literature Review and Research Agenda"*. Telecommunications Policy, 32(8), 531-544

Maícas, J.P., Polo, Y. and Sesé, J. (2009). *"The role of (personal) network effects and switching costs in determining mobile users' choice"*. Journal of Information Technology, 24 (2), 160-171

Merrill Lynch (2010). *Global Wireless Matrix 2009*

OCDE (2007). *OCDE Telecommunications Database*

Steenkamp, J-B, Batra, R., and Alden, D. (2003). *"How perceived brand globalness creates brand value"*. Journal of International Business Studies, 34 (1), 53-65

World Bank Group (2010). *World Development Indicators Online*

In: Mobile Phones: Technology, Networks and User Issues ISBN: 978-61209-247-8
Editor: Micaela C. Barnes et al., pp. 183-189 ©2011 Nova Science Publishers, Inc.

Chapter 6

M-HEALTHCARE: COMBINING HEALTHCARE, HEALTH MANAGEMENT, AND THE SOCIAL SUPPORT OF THE VIRTUAL COMMUNITY

Wen-Yuan Jen[*]
Associate Professor, Institute of Information and Society,
National United University, Taiwan

ABSTRACT

Mobile health services (m-health services) offered through mobile technology have contributed greatly to improved healthcare. At present, most m-health services focus on patient care and treatment; unfortunately, the potential benefit of mobile technology to preventive medicine has not yet been adequately explored. As the population ages and instances of obesity and obesity-related illness among the general population increase, preventative medicine services delivered by m-health systems may help relieve pressure on limited healthcare resources. M-health services are available 24/7 and, more importantly for preventative medicine, have the potential to create virtual support groups that will improve preventative medicine effectiveness. This chapter describes the potential of m-health technology to deliver preventative medicine information and assistance and to create effective support groups among individuals with common health concerns.

Keywords: healthcare, health management, mobile services.

1. INTRODUCTION

Despite the recent global economic slump, the number of mobile phone subscriptions worldwide has reached 4.6 billion and is expected to increase to 5 billion this year (CBSNews, 2010). As mobile phone technology improves, even low-end mobile phones now

---

[*] E-mail: wenyuan.jen@gmail.com

provide banking, health, and other services via text messaging applications. Because of their ubiquitous nature, cellular phones may soon surpass desktop computers in frequency of Internet access.

In the past few years, the healthcare industry has welcomed increased mobile services such as mobile diabetes management services (Ferrer-Roca, 2004), mobile diet and weight reminder services (Jen, 2010; Jen 2009), RFID emergency room patient services (Chao and Jen, 2009), and mobile patient monitoring services (Li, et al., 2008). While most current mobile healthcare services focus on treatment, the potential for their use in prevention is great. Khaw et al. (2008) pointed out that healthcare costs for individuals who maintain healthy lifestyles are 49% lower than people with unhealthy lifestyles, and they live up to 14 years longer. An indispensible part of a healthy lifestyle is healthcare management, which is made much more efficient through the use of mobile communication devices.

This article examines preventative mobile health services and the contribution mobile technology might make to the formation and maintenance of the social support networks essential to successful health management.

2. MOBILE HEALTHCARE

Healthcare services encompass medicine, dentistry, nursing, pharmacy, and the allied health fields. Mobile healthcare (m-healthcare or m-health) integrates the technologies of mobile telecommunication and multimedia in healthcare services delivery (Istepanian, et al., 2005). Delivering healthcare services via mobile devices (e.g. handheld PCs, PDAs, etc.) is a convenient way to help people manage their personal health – whether they are patients currently under a physician's care or healthy individuals committed to staying well.

2.1. Illness Treatment – Mobile Healthcare Services

According to United Nations data, more than one million people turn 60 every month worldwide. The number of people aged 65 and over will double as a proportion of the global population from 7% in 2000 to 16% in 2050. This will be the first time on record that the aged outnumber children below 14 years of age. In Japan, the number of those over 65 years old reached 21.6% in 2008, meaning that Japan has become a Hyper-aged society (United Nations, 2010). In many countries, increasing percentages of elderly and decreasing birth rates mean that society will face increased economic difficulties as the wage-earning population shrinks and the healthcare needs of an ageing population expand.

Another strain on healthcare resources is the "obesity epidemic." This health problem affects young and old alike as young people, because of poor eating habits, are developing health problems that used to affect only adults, such as high blood pressure, high cholesterol, Type 2 diabetes, and heart disease.

M-healthcare systems currently provide help in treating illness. Out-patients under a physician's care, for example, maintain contact with their physician or hospital and can send information and data electronically to their healthcare provider. This data is accumulated in the individual patient's record, which is regularly checked by their physician. If any irregularity is diagnosed, the doctor can advise patients via cell phone or email, or ask patients to schedule an office visit. If an emergency occurs, individuals can also use their cell phones to alert their family or medical services providers. Employing m-healthcare services increases patient freedom by allowing them to recover at home. It also decreases the need for in-patient services and alleviates the worry working family members feel for their home-bound elderly.

Figure 1 illustrates how m-healthcare services facilitate communication and data sharing between participating individuals, their healthcare providers, and concerned family members.

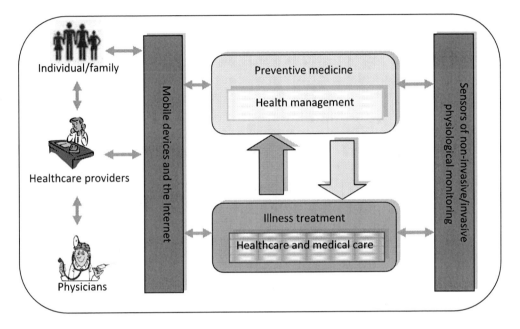

Figure 1. A diagram of the illness treatment and illness prevention.

2.2. Illness Prevention – Mobile Health Management Services

While no one disputes the value of effective health care, of even greater value is the maintenance of a healthy condition and the avoidance of illness or injury. While "health management" includes activities related to the development and implementation of policies governing services designed to improve health, it also applies to the individual (Hunter and Brown, 2007). Health management is a habit, an attitude, and a requirement of daily life, and, as people are now living longer, they should develop health management skills as early as possible.

Wireless communication technologies, including Bluetooth, Cellular, Infrared, UWB, Wi-Fi, and WiMAX, have been successfully applied in the healthcare industry. The functions of mobile health management include "lifestyle management" and "chronic disease management."

  1) Lifestyle management. The seven leading health risk factors, alcohol consumption, high blood cholesterol, obesity, high blood pressure, smoking, low

fruit and vegetable intake, and physical inactivity account for nearly 60% of the disease burden in Europe (WHO, 2005). There is no doubt that the influence of these factors could be mitigated by individual health self-management. Health self-management is the management of day-to-day health-related behaviors and decisions by the individual (Lorig and Holman, 2003). There are many m-health products and services currently available that assist in health self-management, including Bio-shirts (that record the individual's pulse and body temperature), mobile fitness services (for health and physical strength measurement), mobile diet services (that provide nutritional information about thousands of food items), and mobile health diaries (which allow participants to enter personal health information for direct transmission to a healthcare provider and which are accessible 24/7 by healthcare professionals).

2) Chronic disease management. Patients with long-term medical conditions employ m-health systems to provide their physician with up-to-date data concerning their blood pressure, blood sugar, diet/exercise program, and caloric intake. Patients can also receive physician-directed reminders and advice. Recent successful applications of m-health systems include programs that monitor hypertension and diabetes. Patients in these programs measure their blood pressure or glucose level with an electronic sensor which, when connected to a cell phone, transfers results directly to the dedicated healthcare service provider. This data is stored in the individual patient's record and is regularly monitored by physicians who follow up with information or instruction.

3. SOCIAL SUPPORT

3.1. Social Support and the Virtual Community

The concept of social support and its relation to health outcomes has been a focus of study in psychosocial epidemiology for years (Uchino, 2005). Low levels of social support have been shown to increase the risk of developing a major depression over a lifespan (Slavich et al., 2009). The positive effects of social support have also been demonstrated. Cohen and Willis (1985) pointed out that social experience is essential to well-being for everyone, from children to older adults. Those people who have strong social support tend to have higher chance of being healthy and lower rate of disease and early death. In another study, Kayman, et al. (1990) found that over 70% of successful weight losers received social support. Because healthcare or health management programs are long-term activities, some participants may feel frustrated in the short term. With the support of a concerned community, they may be more inclined to continue with programs they have joined.

In traditional support groups, people with common concerns gather to share experiences, ask questions, and even provide emotional support. Now, in the age of the Internet and wireless communication, "virtual" support groups, virtual communities (VC), can make a significant contribution to this social support. A virtual community (VC) requires (1) people who interact socially with each other in a community; (2) a purpose providing a reason such as an interest, need, information exchange, or service, for users to participate in the community; (3) policies for rules, protocols, and laws to guide user behavior; and (4) computer systems to support and mediate social interaction as well as assist in generating a sense of togetherness (Preece, 2000).

In the area of health care, VC is defined as mental health and social support interventions that have the function and character of self support groups (Eysenbach, 2004). An illustration of the value of social support to health management appears in Laitinen and Sovio (2002), which found that the support of a close community was important to successful weight maintenance or weight loss. Figure 2 shows how a weight management VC might help people find the strong social support they will need to achieve their weight management goals. The VC employs web-based and cell phone communications to link its users into a tight community (Jen, 2010). The social support users find in VCs motivates them to participate in health management programs and activities.

3.2. Mobile Virtual Communities and Health

The dramatic increase in the number of people willing to share information via mobile devices has given rise to a concomitant growth in Mobile Virtual Communities (MVC). Through MVCs, social support and interpersonal interaction is possible anytime, anywhere. Live chat, notification, and subscription to member information functions of the MVC promote social interaction in one-to-one, one-to-many, or many-to-many formats. Mobile virtual communities have been employed in the telemedicine domain to improve the effectiveness of the basic healthcare system (van Beijnum, 2009), and healthcare professionals, patients, informal caregivers, patient family members, and the general public all enjoy the benefits of MVC participation (Demiris, 2006). The MVC acts, indeed, as an indirect healthcare "professional," helping patients to recover from an illness and assisting those in good health to maintain their current condition.

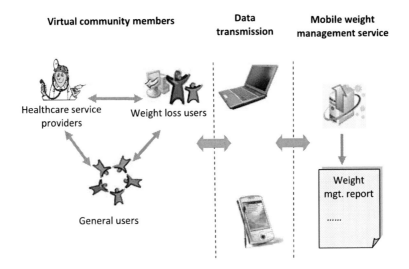

Figure 2. The architecture of the mobile weight management service of VC (Jen, 2010).

CONCLUSION

Once an information-sharing tool for medical professionals in treating patients, mobile technology is finding greater use among the general population as an aid to preventative medicine. M-healthcare services such as health lifestyle management, fitness services, and medical information/health education services are becoming main stream. In developing countries as well, the growth of mobile phone promises to bring m-healthcare to those in greatest need. A full 64% of all mobile phone users can now be found in the developing world (United Nations Department of Economic and social Affairs, 2007), and as these numbers grow, so will the numbers of people gaining access to medical care, health education, and community support.

This paper describes the promise mobile technology holds for integrating healthcare services and empowering patients; the task that now lies before us is to empirical demonstrate that promise in clinical outcomes.

REFERENCES

CBSNews. *Number of Cell Phones Worldwide Hits 4.6B. 2010/8/7.* Available from: URL: http://www.cbsnews.com/stories/2010/02/15/business/main6209772.shtml.

Chao, C.C. and Jen, W.Y. *Patient safety management: RFID technology to improve Emergency Room medical care quality.* In T. Shih and Q. Li (Eds.), Ubiquitous Multimedia Computing. CRC Press, Taylor and Francis Company, 2009; 323-339

Cohen, S. and Willis, T.A. (1985). *Stress, social support, and the buffering hypothesis.* Psychological Bulletin, 98, 310-357.

Demiris, G. (2006). *The diffusion of virtual communities in health care: concept and challenges.* Patient Education and Counseling, 62, 178-188.

Eysenbach, G., Powell, J., Englesakis, M., Rizo, C. and Stern, A. (2004). *Health related virtual communities and electronic support groups: systematic review of the effects of online peer-to-peer interactions*, BMJ, 328, 1166-70.

Ferrer-Roca, O., Cárdenas, A., Diaz-Cardama, A. and Pulido, P. (2004). *Mobile phone text messaging in the management of diabetes.* Journal of Telemedicine and Telecare, 10, 282-285.

Hunter D.J. and Brown, J.A. (2007). *Review of health management research.* Europe Journal Public Health, 17, 33–37.

Istepanian R.S.H., Laxminarayan, S., and Pattichis, C. S., eds. M-Health: *Emerging Mobile Health Systems.* Springer: 2005. New York: Kluwer/Plenum.

Jen, W.Y. (2009). *Mobile healthcare services in school-based health center*, International Journal of Medical Informatics, 78, 425-434.

Jen, W.Y. (2010). *The adoption of mobile weight management service in a virtual community: The perspective of college students.* Telemedicine and e-Health, 16, 490-497.

Kayman, S., Bruvold, W. and Stern, J.S. (1990). *Maintenance and relapse after weight loss in women: Behavioral aspects.* American Journal Clinical Nutrition, 52, 800-807.

Khaw, K.T., Wareham, N., Bingham, S., Welch, A., Luben, R. and Day, N. (2008). *Combined impact of health behaviours and mortality in men and women*: the EPIC-Norfolk prospective population study. PLoS Medicine, 5, e70.

Laitinen, J., Ek, E. and Sovio, U. (2002). *Stress-related eating and drinking behavior and body mass index and predictors of this behavior*. Preventive Medicine, 34, 29-39.

Li, B.N., Fu, B.B. and Dong, M.C. (2008). *Development of a mobile pulse waveform analyzer for cardiovascular health monitoring*. Computers in Biology and Medicine, 38, 438-445.

Lorig, K.R. and Holman, H.R. (2003). *Self-management education: history, definition, outcomes, and mechanisms*. Annals of Behavioral Medicine, 26, 1-7.

Mattila, E. (2010). *Design and evaluation of a mobile phone diary for personal health management*. VTT Publication, 742, 1-81.

Preece, J. *Online communities: designing usability, supporting sociability*. Chinchester: John Wiley and Sons, 2000.

Slavich, G.M., Thornton, T., Torres, L.D., Monroe, S.M., Gotlib, I.H. (2009). *Targeted rejection predicts hastened onset of major depression*. Journal of Social and Clinical Psychology, 28, 223-243.

Uchino, B.N. (2005). *Social support and physical health: understanding the health consequences of relationships*. American Journal of Epidemiology, 161, 297-298.

United Nations Department of Economic and Social Affairs (2007). *Division for Public Administration and Development Management, Compendium of ICT Applications on Electronic Government - Volume 1*. Mobile Applications on Health and Learning, New York: United Nations.

United Nations, Aging society. 2010/8/7. *Available from:* URL:http://wisdom.unu.edu/en/ageing-societies/.

van Beijnum, B.J.F., Pawar, P., Dulawan, C.B. and Hermens, H.J. (2009): *Mobile Virtual Communities for Telemedicine: Research Challenges and Opportunities*. International Journal of Computer Science and Applications, 6, 19-37.

WHO 2005. *The European health report 2005*. Public health action for healthier children and populations. Copenhagen, Denmark: WHO. 129.

In: Mobile Phones: Technology, Networks and User Issues   ISBN: 978-1-61209-247-8
Editors: Micaela C. Barnes et al., pp. 191-225   © 2011 Nova Science Publishers, Inc.

Chapter 7

OLS - OPPORTUNISTIC LOCALIZATION SYSTEM FOR SMART PHONES DEVICES

Maarten Weyn[*]
Artesis University College of Antwerp - Belgium
Martin Klepal[†]
Cork Institute of Technology - Ireland

Abstract

People are eager to locate their peers and stay connected with them. The Opportunistic Localization System (OLS) makes that simple. OLS's core technology is an opportunistic location system based on smart phone devices. One main advantage is that it seamlessly works throughout heterogeneous environments including indoors as opposed to GPS based systems often available only outdoors under the unobstructed sky.

OLS is a phone-centric localization system which grasps any location related information that is readily available in the mobile phone. In contrast to most of the competing indoor localization systems, OLS does not require a fixed dedicated infrastructure to be installed in the environment, making OLS a truly ubiquitous localization service. The latest version of OLS strives to reduce the system ownership cost by adopting a patent covered self-calibration mechanism minimizing the system installation and maintenance cost even further.

OLS's architecture migrated from the original client-server to the current service oriented architecture to cope with the increasing demands on reusability across various environments and platforms. Thus, to scale up service for a large variety of clients. The location related information used for the estimation of mobile device location are used by existing signals in the environment and can be sensed by a smart phone. The readily available information, depending on the phone capability, is typically a subset or all of the following: the GSM/UMTS signal strength, WiFi signal strength, GPS, reading from embedded accelerometers and Bluetooth proximity information.

The reliability and availability of input information depends strongly on the actual character of the mobile client physical environment. When the client is outdoors

---

[*]E-mail address: maarten.weyn@artesis.be
[†]E-mail address: mklepal@cit.ie

under the unobstructed sky, GPS is a favorable choice typically combined with pedometer data derived from the 3D acceleration and compass measurement. When the client is in a dense urban and indoor environment the GSM/UMTS and WiFi signal strength combined with pedometer data usually performs best. If indoor floor plan layouts are available, the map filtering algorithm can further contribute to the location estimation. However, the information about the actual type of environment is not available and proper importance weighting of all input information in the fusion engine is paramount. This chapter will describe how OLS developed and particularly give insight in OLS core technology, which is the adaptive fusion of location related information that is readily available in smart phones.

The experiments carried out, shows when considering all location related information for the object location estimation namely the GSM and WiFi signal strength, GPS, PDR and map filtering, the presented opportunistic localization system achieves a mean error of 2.73 m and a correct floor detection of 93% in common environments. The achieved accuracy and robustness of the system should be sufficient for most of location aware services and application, therefore having a potential of enabling truly ubiquitous location aware computing.

PACS 89.20.Ff

Keywords: Opportunistic Localization, Positioning, Sensor Fusion, Smart Phone, Bayesian Estimation, Pattern Matching

AMS Subject Classification: 47N30

1. Introduction

The rapid proliferation of social networking websites such as Facebook, Twitter, MySpace, Bebo, etc., demonstrate that people are eager to stay connected with their peers and are open to sharing personal information with others. Hence, this makes it an attractive proposition to demonstrate the capabilities of OLS services. A number of geo-social networking solutions exist where location information enhances the social interaction between users but these are mostly limited to outdoor environments. OLS can bridge the gap between outdoor and indoor localization, OLS could become another case of campus culture having a major impact on the real world, like Facebook or YouTube.

There also exists a number of health and safety applications that can benefit from indoor location information such as navigation of emergency response, first responder, fire officers, and tracking in industrial environment to name a few. Security systems can also benefit greatly from the availability of location information, for example tracking contract staff on a large industrial site, ensure an employee is in the place they are meant to be. OLS services provides this capability to the developers of applications for occupancy safety and security with an easy-to-access application programming interface, the applications can become location aware almost immediately therefore extending their product offering.

The future of localization systems most likely will evolve towards systems which can adapt and cope with any available information provided by mobile clients. However, one of the common disadvantages of many existing localization systems is the need for dedicated devices and proprietary infrastructure in the operation area of the indoor localization system.

The increasing proliferation of mobile devices, such as Personal Digital Assistants (PDAs) and smart phones, has fostered growing interest in location based applications that can take advantage of available information in the environment and can be extracted by the mobile device. This type of localization is called opportunistic localization. For example, most mobile devices can provide GSM related data like the connected cell tower identification and signal strength, whereas more advanced devices are equipped with WiFi, GPS or a combination of the previous. More recent mobile devices also have inertial sensors build in (mostly accelerometers and or compasses) which can be used as extra localization information using the Pedestrian Dead Reckoning (PDR) [1] principle. PDR uses the internal accelerations to estimate the person distance advances.

There has been conducted significant amount of research in the area of localization data fusion and hybrid localization systems for decades. Some authors refer to the systems that combine location data from different dedicated localization systems to be the opportunist systems. Other authors restrict and narrow the use of term *opportunistic* only to the localization systems which seize the opportunity and take advantage of any readily available location related information in the environment that can supply a network for a mobile device without relying on installation of any dedicated hardware. This is believed to lead towards a truly ubiquitous availability of localization services. In this chapter, the term opportunistic refers strictly to the later narrower meaning and to prevent any possible confusion we define here the opportunistic system as follows:

Definition *"The opportunistic localization system is a system that seizes the opportunity and takes advantage of any readily available location related information in an environment, to supply a network for a mobile device and allows an estimation of the mobile device absolute or relative position without relying on installation of any dedicated localization hardware infrastructure."*

The rest of the chapter will describe the architecture of the present OLS system in Section 3., together with different sensors which can be used to obtain location aware information readily available on the smart phone. Their combination in the Bayesian recursive fusion engine is explained in Section 4., with two main parts, the motion and measurements model. In Section 5. the influence of a predicted and measured fingerprint on the accuracy of an RSS-based indoor location system is discussed. The fusion and the fingerprint predictions are discussed using experiments in Section 6. and conclusions are made in Section 7..

2. Current State-of-the-Art

As shown by [2], a lot of techniques and technologies can be used for localization. One of these technologies, other than GPS, is GSM or the cellular network which is widely used for the localization of handheld devices [3]. Also, systems using WiFi like [4–6], have already proven themselves. Since these systems strongly depend on the WiFi access point placement and an extensive signal propagation measurement called the RF-fingerprint, they are not considering other location related inputs.

PlaceLab [7] combines technologies such as GPS, WiFi, GSM, and Bluetooth beacons. Whereas commercial systems like Skyhook Wireless [8, 9] combines GSM and WiFi beacons to augment or replace GPS. They call this hybrid localization. And Rosum [10, 11] augments GPS by using TV signals.

Sensor data fusion was already discussed before by Hall *et al.* [12–14] but was used mostly in the field of robotics [15–17]

The opportunistic use of data was already shown in [18–20] where data is exchanged between nodes when the opportunity arises. The term opportunistic localization is also used by [21], to only localize a mining vehicle when there is the opportunity; in their case, when the vehicle passes a known or deductible recognition point. In [22], opportunistic visual observations are used for localization and in [23] GPS and GLONASS signals are used opportunistically.

In OLS, GSM is not used for cell ID localization as in PlaceLab, but the signal strength measurements are directly used in the measurement model, similar to OLS's WiFi implementation. Accelerometers are already used in other systems like [1, 24], but all other research on PDR mostly uses dedicated accelerometers, whereas OLS uses the built-in sensors of smart phones. This also means that the algorithm should cope with varying placement of the smart phone which may be in a pocket of trousers, with an unknown orientation, for example; where most other systems uses accelerometers placed on a predefined part of the body (foot, chest, shoulder, etc.). OLS focuses on opportunistic sensor data fusion to empower localization, using a adaptive motion and measurement model a Bayesian recursive fusion engine.

3. Architecture

To cope with increasing demands on reusability across various environments and platforms and to scale up service a large variety of clients, OLS's architecture migrated from the original client-server to the current service-oriented architecture. The server side or server cloud runs the main OLS services such as the localization, management, registration, communication service, data fusion engine, and database. The architecture is shown in Figure 1.

In this section, the architecture of OLS is discussed. We will focus on some server side system components and elaborate on different device sensors.

3.1. System Components

The *OLS Server* (or server cloud) is responsible for hosting the application and providing all the services. The *Communication Service* is responsible for providing and maintaining a communication interface between the clients and the *OLS Server*. The communication makes use of the LocON data protocol [25]. This enables a transparent and standardized communication between the client and the server. The *Registration Service* registers localized objects for OLS clients and the users of the *Location Service*. The *Location Service* provides location updates for localized objects and instantiates and manages the instances of the main processing components of the *OLS Server*. The *Data Fusion Engine* provides the data fusion service, which will be explained in Section 4.. The *Management Service* manages and maintains the *OLS Server* and facilitates its administration. The *Database* stores environment descriptions such as maps, fingerprints, and floor transition areas. It also stores information necessary for the data fusion algorithm. The *Mobile Client* is a smart phone

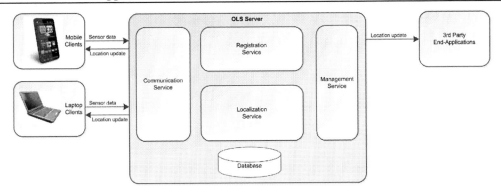

Figure 1. The System Architecture.

device, which is localized by the *OLS server* using the streamed sensory data from its GPS, GSM/UMTS, accelerometers, WiFi, and Bluetooth, compasses and other available sensors. The *Mobile Client* can also subscribe for location updates in order to further process and visualize its location. The *Laptop Client* is a third-party client typically with higher processing and visualization capabilities. The *Laptop Clients* can run third-parties location based applications, that further process the location data provided by OLS. The *Communication* and *Localization Service* are described in the next sections.

3.1.1. Communication Service

One of the tasks of the *Communication Service* is to translate the sensor data coming from the devices to messages which can be used by the *Location Service*. To allow opportunistic localization, it is important to support a wide variation of devices. For this reason, a strict data communication protocol developed in FP7 EU project called LocON is used [26]. The LocON data protocol defines the data format, which in OLS, can be used in TCP or UDP messages. The data carried by the protocol can contain different kind of messages. The three types used in OLS are *Position Messages*, *Extended Data Messages*, and *Configuration Messages*.

- *Position Messages* are messages containing a GPS or local Cartesian coordinate. In OLS, this type of message is used to send GPS data from a client to the server.

- *Extended Data Messages* are messages that can contain any sensor data, formatted in a predefined way. All sensor data from clients is send to the *OLS Server* using this type of messages.

- *Configuration Messages* are normally send from the server to the device to adapt configuration settings of the devices, for example the update rate of WiFi signal strength measurements.

3.1.2. Localization Service

The *Localization Service* is responsible for processing the data coming from the devices through the *Communication Service* and estimate the position of each device. The different

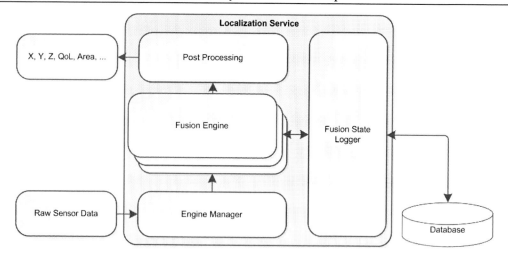

Figure 2. Localization Service.

components of the *Localization Service* are shown in Figure 2.

The raw sensor data is sent to the *Localization Service* and is received by the *Engine Manager*. This component is responsible for dividing the data across different *Fusion Engines* which can be located on different hardware components. This design makes it possible to scale the localization service according to the number of devices which have to be supported by the system. The *Fusion Engines* themselves are stateless. The fusion state (posterior fusion state, fusion engine settings, etc.) are stored in the *Fusion State Logger*. This logger persists all the data to the database, but also keeps the most recent states in memory to limit additional delays. The working of the *Fusion Engines* is explained in Section 4.. The *Post Processing* uses the posterior probability of the object state estimated by the *Fusion Engine*, to calculate the estimated most likely location (x, y, z) and its quality factor called *Quality of Location* (QoL) represents the mean error of the estimated location. Furthermore the *Post Processing* component also extracts additional information from the object state such as, speed and direction of the device and the area name of the object containment.

3.2. Device Sensors

This section describes the sensors used by the OLS system. The selection of the sensors depends on the availability of environmental signals and types of sensors supported by the hardware of the device.

3.2.1. Global Navigation Satellite System

Satellite-based navigation is the leading technology for outdoor navigation. GPS has already exhaustively proven itself and with the introduction of its European counterpart Galileo, GNSS services will only be extended and improved. Nevertheless, GNSS cannot be used as the only positioning technology to meet all requirements in all types of terrain. Although possible in some situations [27], GNSS signals cannot penetrate sufficiently most indoor environments to be used by a normal receiver. In urban environments and other RF shadowed environments, satellite navigation usually performs poorly. Moreover, the Time to First Fix (TTFF), a cold start of a GPS device can take up to a few minutes [28], which is far too long for many applications.

3.2.2. WiFi

Localization using WLAN, most notoriously based on IEEE802.11x standard commercially branded as WiFi, is probably the most commonly used technology for indoor localization as it benefits from the omnipresence of WiFi networks. *Nearest Access Point* (AP), *Multi-lateration* [29], and *Pattern Matching* [30] are techniques for pinpointing a mobile node position. Since the idea of OLS is to create sensors which can be re-used as much as possible, *Multi-lateration* using timing is rather difficult to achieve since it is not feasible to get precise timing information from standard of-the-shelf devices in a generic way. This makes time dependent localization techniques like *Time Difference of Arrival* (TDOA), *Time of Arrival* (TOA) and *Round Trip Time* (RTT) not possible to use for generic PDAs in a phone-centric localization system. In OLS fingerprinting based signal strength is implemented, this requires a fingerprint database, which calibration can be streamlined and automated by a recently developed technique known as the fingerprint self-calibration [31].

3.2.3. GSM

Every mobile phone has the identification and signal strength of the cell tower to which it is connected (the serving cell). A broadcast message of this tower informs all connected cell phones about maximum eight neighboring cell towers that can be used in case a handover is needed. The measured GSM/UMTS signals strengths from this serving and up to eight neighboring cells are treated similarly to WiFi signal strength, as discussed earlier. Cell-ID localization is used in different case studies [32, 33] and applications already. In Cell-ID localization the mobile phone is located in the area covered by the connected cell tower. The implementation of Cell-ID localization adds straightforward coarse-grained indoor and outdoor location information to OLS, but the location of all cell towers is needed. However, the main benefit of GSM is in urban (near-) indoor environments where the pattern, created from the GSM field strengths of all cell towers, can provide localization where GPS fails due to the blockage of the satellite signals (shadowing), as similar to [34]. Thanks to the influence of surrounding obstacles such as walls on the GSM signal propagation the fingerprint of neighboring cell towers field strengths differs significantly across and in and out of buildings. These signal strength spatial variations are harnessed for localization purposes similarly to WLAN based fingerprinting like in [4]. The difference in OLS towards other research on indoor GSM localization like [34], is that a standard PDA does not allow getting system information of all surrounding cell towers. Since we try to build OLS as generic as possible that only use the information which is readily available in most mobile phone. Therefore, OLS uses a standard GSM fingerprinting to cover areas with poor GPS and WiFi coverage. The main strength of OLS lies in the combination of all complementing technologies providing ubiquitous localization service across heterogeneous environments and signal quality coverage.

3.2.4. Accelerometer

The accelerometers are used to process step-detection as part of an adapted *Pedestrian Dead Reckoning* PDR algorithm as described in [35]; the number of steps will be send to the server, together with an estimated distance. The accelerometers provide also extra information about the state of the person using the device, such as standing still, moving or climbing stairs, which is used in the motion model described next, if the step detection cannot be performed due to an irregular motion pattern of the person during a chat in the corridor for example.

3.2.5. Compass

Compass data is used to detect the absolute orientation of a person if the phone placement on the person is known and constant. Otherwise only relative phone orientation changes detected by the compass are considered in the person motion model. This is shown in Algorithm 1.

Algorithm 1 Compass measurement $\alpha^t_{compass}$

1: **if** orientation of phone to person $\Delta\alpha^t_{person}$ known
   and orientation of environment $\alpha_{environment}$ known **then**
2: $\quad \alpha^t_{device} = \alpha^t_{compass} - \Delta\alpha^t_{person} - \alpha_{environment}$
3: $\quad$ **return** α^t_{device}
4: **else**
5: $\quad \Delta\alpha^t_{device} = \alpha^t_{compass} - \alpha^{t-1}_{compass}$
6: $\quad$ **return** $\Delta\alpha^t_{device}$
7: **end if**

3.2.6. Bluetooth

There have been experiments with Bluetooth localization [36] similar to WiFi localization but this requires a dedicated infrastructure that OLS tries to avoid. Therefore, the main purpose of a Bluetooth is device proximity detection [37] and possibly also 'Object Binding' [38].

3.2.7. Other Contextual Information

Other user activity, such as measuring the last time of a keystroke, diary information stored in the smart phone and any additional contextual location related information can be taken into account in the OLS system. However, this is not further elaborated in this chapter.

4. Seamless Sensor Fusion

This section will explain how the data coming from these mentioned sensors can be combined for localization purposes. This processing is handled in the *Fusion Engine*.

The *Fusion Engine* runs a nonlinear recursive Bayesian filter to seamlessly combine all incoming location information. A particle filter [39], a technique which implements the recursive Bayesian filtering using the sequential Monte Carlo method, is currently one of the most advanced techniques for sensor data fusion. A particle filter allows modeling the physical characteristics of the tracked object movement and environment. This motion model describes the possible transition from the previous object to the next state by using first and second order parameters of the object kinematic model such as the position, speed, heading, etc. By doing so it is able to incorporate also environmental constraints like walls and other obstacles to remove impossible objects motion trajectories. The noisy measurements (observations) like WiFi or GSM signal strength or GPS coordinates can be modeled in different ways using the measurement model.

A particle filter is based on a set of random discrete samples with a weight, or particles, to represent the probability density of the object state. The amount of particles used is dependent on the available processing power and needs of the application concerning speed and accuracy. The number of particles is automatically adapted based on the KLD sampling algorithm by [40, 41].

The measurement model defines a weight of every particle corresponding to the probability of the state of the particle with that certain measurement. This weight will be used during the resampling step of the particle filter process. Since the measurement model of a certain technology gives a weight (conforms its probability) to every particle, fusion of different technologies can be done by multiplying the normalized weights of the different measurement models of every particle. After each update step the particles are resampled according to their weight. The motion model can also be adapted depending on the knowledge coming from sensors, going from random Gaussian samples to specific controlled transitions. A possible implementation with the sensors mentioned earlier is described in [42]

The main advantage of particle filters towards other Bayesian filtering like Kalman Filters, for example, is that it can deal with non-linear and non-Gaussian estimation problems. Additionally, the particle filter through the particles motion model, can consider environment constraints known as the map filtering, where environmental knowledge like the floor plan layout is used to remove impossible trajectories [4].

4.1. Particle Filter Algorithm

The particle filter directly estimates the posterior probability of the state $\mathbf{x}_t$, which is expressed with the following equation [43]:

$$p(\mathbf{x}_t|\mathbf{z}_t) \approx \sum_{i=1}^{N} w_t^i \delta(\mathbf{x}_t - \mathbf{x}_t^i) \qquad (1)$$

where $\mathbf{x}_t^i$ is the i-th sampling point or particle of the posterior probability with $1 \leq i \leq N$ and w_t^i is the weight of the particle. N represents the number of particles in the particle set and $\mathbf{z}_t$ represents the measurement.

Algorithm 2 describes the generic algorithm of a particle filter. The input of the algorithm is the previous set of the particles X_{t-1}, and the current measurement $\mathbf{z}_t$, whereas the output is the new particle set X_t.

In our OLS system, the state **x** of a particle represents the position (x_t^i, y_t^i, z_t^i), its velocity v_t^i, heading α_t^i and validity $valid_t^i$ (for example, to identify if a particle has crossed a wall).

Algorithm 2 Particle_Filter ($\mathcal{X}_{t-1}, \mathbf{z}_t$)

1: $\bar{\mathcal{X}}_t = \mathcal{X}_t = \emptyset$
2: **for** $i = 1$ to N **do**
3:    sample $\mathbf{x}_t^i \sim p(\mathbf{x}_t|\mathbf{x}_{t-1}^i)$
4:    assign particle weight $w_t^i = p(\mathbf{z}_t|\mathbf{x}_t^i)$
5: **end for**
6: calculate total weight $k = \sum_{i=1}^{N} w_t^i$
7: **for** $i = 1$ to N **do**
8:    normalize $w_t^i = k^{-1} w_t^i$
9:    $\bar{\mathcal{X}}_t = \bar{\mathcal{X}}_t + \{\mathbf{x}_t^i, w_t^i\}$
10: **end for**
11: $\mathcal{X}_t = $ **Resample** ($\bar{\mathcal{X}}_t$)
12: **return** $\mathcal{X}_t$

The algorithm will process every particle $\mathbf{x}_{t-1}^i$ from the input particle set $\mathcal{X}_{t-1}$ as follows:

1. Line 3 shows the prediction stage of the filter. The particle $\mathbf{x}_t^i$ is sampled from the transition distribution $p(\mathbf{x}_t|\mathbf{x})$. The set of particles resulting from this step has a distribution according to (denoted by $\sim$) the prior probability $p(\mathbf{x}_t|\mathbf{x}_{t-1})$. This distribution is represented by the motion model.

2. Line 4 describes the incorporation of the measurement $\mathbf{z}_t$ into the particle. It calculates for each particle $\mathbf{x}_t^i$ the *importance factor* or *weight* w_t^i. The weight is the probability of the received measurement $\mathbf{z}_t$ for particle $\mathbf{x}_t^i$ or $p(\mathbf{z}_t|\mathbf{x}_t)$. This is represented by the measurement model.

3. Line 7 through 10 are the steps to normalize the weight of the particles. The result is the set of particles $\bar{\mathcal{X}}_t$, which is an approximation of posterior distribution $p(\mathbf{x}_t|\mathbf{z}_t)$.

4. Line 11 describes the step known as *resampling* or *importance resampling*. After the resampling step, the particle set that is previously distributed equivalent to the prior distribution $p(\mathbf{x}_t|\mathbf{x}_{t-1}, \mathbf{z}_{t-1})$ will be changed to the particle set $\mathcal{X}_t$ which is distributed in proportion to $p(\mathbf{x}_t|\mathbf{x}_{t-1}, \mathbf{z}_t)$.

This adaptive character allows a particle filter to switch seamlessly between different measurement and motion models depending on the data coming from the mobile device. On top of the location we can also infer a quality of location parameter $Qol(x)$. If we receive data that we can use to infer an accurate location, the particles will group closely together around the real location, if we have very noisy or coarse grained information, the particles will spread towards all possible locations, keeping into account the measurement and the previous location using the motion model. This will also allow us to use besides the estimated location, a good Quality of Location parameter gives an idea of the accuracy of the location:

$$Qol = E\left(\left|\vec{x}_t - \vec{x}'_t w^i_t\right|\right) = \frac{1}{N_p}\left(\left|\vec{x}_t - \vec{x}'_t w^i_t\right|\right) \tag{2}$$

4.2. Motion Model

The motion model or transition model describes the possible transition from a state $t-1$ to a state at t. The motion model depends on the tracked device type. A person will have a different motion model then a car for example. In this section, we will describe the motion model of a person.

4.2.1. Person Motion Model

Let $\mathbf{x}^i_t$ denote the state vector that describes a particle position representing a person in a local Cartesian coordinate. The motion of the particle can be modeled with:

$$\mathbf{x}^i_t = \mathbf{x}^i_{t-1} + \mathbf{v}_t \Delta t \cos(\alpha_t) + \eta_t \tag{3}$$

Where $\mathbf{v}_t$ denotes the velocity vector and α_t the heading vector at time t. η_t is Gaussian distributed noise called roughening or jittering [44]. Roughening is adding a jitter to mitigate sample impoverishment [45].

$\mathbf{v}_t$ and α_t are calculated as follows:

$$\begin{aligned} \mathbf{v}_t &= N\left(\mathbf{v}_{t-1}, \sigma_v^2\right), \text{ where } \sigma_v = \sqrt{\Delta_t} \\ \alpha_t &= N\left(\alpha_{t-1}, \sigma_\alpha^2\right), \text{ where } \sigma_\alpha = \frac{\pi}{2} - \frac{\arctan\left(\frac{\sqrt{v_t}}{2}\right)}{\sqrt{\Delta_t}} \end{aligned} \tag{4}$$

This means that the velocity vector is defined by the previous velocity and the time. More specific, the new velocity vector is randomly chosen from a Gaussian distribution with the mean value of the previous velocity and the standard deviation equal to the square root of time the time between the previous and current update. Similarly, the standard deviation of the direction is depending on the speed and the time between the previous and current update as shown in Figure 5. The slower a person moves the bigger the chance he or she will change direction. An example is shown in Figure 3 where the black dot represents the initial position of the person and the crosses are all the particles after the motion update of 1 s. Figure 3(a) shows the prediction step with an initial speed of 3 m/s and Figure 3(b) shows the same prediction step, but with an initial speed of 5 m/s. Both initial directions are 0 (right direction on the graph). This is also visualized in Figure 4.

4.2.2. Sensor Inputs

This motion model described here does not incorporate any additional information from the sensors. As soon as sensory data is available to the fusion engine, it is applied in the motion model. An example is the speed and direction. The speed can be estimated through PDR using accelerometers. Authors in [35] use the accelerometer of a mobile phone to estimate the displacement of a person and therefore also the speed. Another way to procure the speed is from a GPS sensor, since this provides secondary estimation of speed and direction

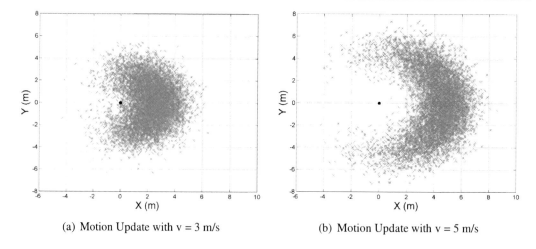

Figure 3. Motion Update with different speeds.

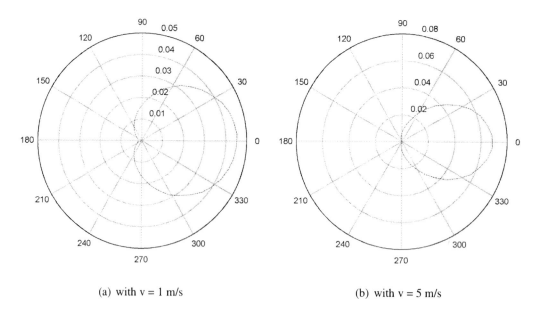

Figure 4. Probability of new direction with initial direction 0 and different speeds.

based on the previous and current measured position. Also, the absolute direction can be estimated by a digital compass.

Accelerometer The accelerometer can provide, through PDR, three different types of information as described in the paper by [35]. First, the state of the object is *'Not Moving'*. Second, the object *'Is Moving'* but the PDR algorithm cannot detect the number of steps taken. And finally, the *'Walking'* state when the algorithm is able to calculate the number and distance of the steps. This will result in the estimation of speed $\mathbf{v}_t$ shown in Algorithm 3, where the speed is sampled from a Gaussian distribution with a mean speed calculated

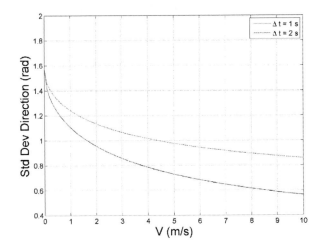

Figure 5. Correlation between Standard Deviation of Direction and Speed.

by PDR and standard deviation of the $\frac{1}{20}$ as the standard deviation which is chosen if no information about the speed is known. This value is based on the error of the measurement and the PDR algorithm [35]. If we know the device is not moving the speed is set to 0 m/s.

Algorithm 3 Speed estimation with PDR input from accelerometers

1: **if** pdr_{moving} **then**
2:    **if** $pdr_{speedknown}$ **then**
3:       $v_t = N(v_{pdr}, \sigma_v^2)$, where $\sigma_v = \frac{1}{20} \cdot \sqrt{\Delta_t}$
4:    **else**
5:       $v_t = N(v_{t-1}, \sigma_v^2)$, where $\sigma_v = 1 \cdot \sqrt{\Delta_t}$
6:    **end if**
7: **else**
8:    $v_t = 0$
9: **end if**

GPS gives an estimation of speed and the direction (0 degrees = Nord). This information is used in the motion model as shown in Algorithm 4.

The speed uses a standard deviation of $\frac{1}{5}$ m/s. This value is dependent on the GPS receiver and should denote the error of the speed measurement in the speed interval of walking speed. For the direction, the direction given by the GPS receiver is geometrically averaged with the previous speed. This is because the direction given by a GPS on walking speed is prone to error.

Compass If we could directly use the direction from the compass we could sample the direction as:

$$\alpha_t = N(\alpha_{compass}, \sigma_{compass}^2)$$

Algorithm 4 Speed and direction estimation with GPS input

1: **if** $gps_{speedknown}$ **then**
2: $\quad v_t = N(v_{gps}, \sigma_v^2)$, where $\sigma_v = \frac{1}{5} \cdot \sqrt{\Delta_t}$
3: **else**
4: $\quad v_t = N(v_{t-1}, \sigma_v^2)$, where $\sigma_v = 1 \cdot \sqrt{\Delta_t}$
5: **end if**
6: **if** $gps_{bearingknown}$ **then**
7: $\quad \alpha_t = N(\sqrt{\alpha_{gps} \cdot \alpha_{t-1}}, \sigma_\alpha^2)$, where $\sigma_\alpha = 0.5\pi - \frac{\arctan\left(\frac{\sqrt{v_{gps}}}{2}\right)}{\sqrt{\Delta_t}}$
8: **else**
9: $\quad \alpha_t = N(\alpha_{t-1}, \sigma_\alpha^2)$, where $\sigma_\alpha = 0.5\pi - \frac{\arctan\left(\frac{\sqrt{v_t}}{2}\right)}{\sqrt{\Delta_t}}$
10: **end if**

Where $\sigma_{compass}$ denotes the standard deviation error of the compass measurements. If we use these equations, we assume that the 0 degree direction of the environment is used to locate the device which is the Nord direction. This can be solved by acquiring a reference direction when calibrating the environment:

$$\alpha_t = N(\alpha_{compass} - \alpha_{reference}, \sigma_{compass}^2)$$

The previous method still assumes that the measured direction of the compass also is the direction of the user, but this also depends on the orientation of the device. One could ask the user to calibrate the direction every time the device is placed in a different position. Yet, this would not be a benefit to the opportunistic concept of OLS. Alternatively, we could use the direction change instead of the absolute direction:

$$\alpha_t = \alpha_{t-1} + N(\Delta\alpha_{compass}, \sigma_{compass}^2), \text{ where } \Delta\alpha_{compass} = \alpha_{compass}^t - \alpha_{compass}^{t-1}$$

4.2.3. Map Filtering

If the environment is known, then the map filtering can be very useful to enhance the accuracy of the location estimation. In the map filtering we remove particles with impossible trajectories of motion. As shown in Figure 6 particles traversing a wall or obstacle will be neglected in the next resampling process.

4.2.4. Transitions

Stairs, escalators, and elevators introduce an additional challenge to be modeled in the motion model. An environment is typically defined as a multiple floor model, where stairs, escalators and elevators form the only possibility to change floors. A stairs can be implemented as a transition line, which triggers the z value of a sample to be changed when crossed. An escalator is similar, but introduces an extra velocity to the particle. An elevator has typically multiple exit floors. This can be implemented by selecting for every sample a random floor. The measurement model (Section 4.3.) will remove the invalid particles in the correction step.

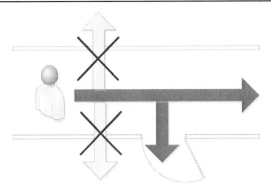

Figure 6. Remove particles with impossible trajectories.

4.2.5. Adaptive Motion Model

All of the mentioned information can be combined in an adaptive motion model, according to Algorithm 5.

4.3. Measurement Model

The measurement model is used to incorporate the sensor information in likelihood function. This function calculates the probability of a sample giving the sensor measurement. This probably is used in the resampling step. For every type of sensor another measurement model is used. Example measurement models of GNNS, WiFi, GSM, and Bluetooth are described here:

4.3.1. GNNS

The likelihood function used for a GNSS measurement (currently GPS is used) takes into account the distance between the sample position (the particle) and the position given by the GNSS sensor (the measurement), the Horizontal Dilution of Precision (HDOP), and the internal error of the sensor. The HDOP is an indication of the horizontal error originating for the GPS satellites geometry. The internal error of a GPS sensor [46] is the error origination from the precision and hardware of the GPS receiver. The coordinates of the measurement is first translated to the local environment Cartesian coordinates. Figure 7 shows the probability for a measurement with Cartesian coordinates $(0,0,0)$ and different HDOPs.

4.3.2. WiFi

In OLS, a pattern matching based localization is used. A new WiFi measurement is compared with the pre-recorded WiFi fingerprint database. This result in a likelihood function which will identify areas on the map with a high correlation between the measurement and the fingerprint, similar to [31]. An example measurement probability for WiFi using this likelihood function can be seen in Figure 8(a).

Algorithm 5 Motion_Model ($\mathbf{x}_{t-1}^i$, $\mathbf{u}_t$, Δ_t)

1: **for** $i = 1$ to *maxRetries* **do**
2:    **Speed**
3:    **if** pdr_{moving} **then**
4:      **if** $pdr_{speedknown}$ **then**
5:         $v_t = N(v_{pdr}, \sigma_v^2)$, where $\sigma_v = \frac{1}{20} \cdot \sqrt{\Delta_t}$
6:      **else**
7:         **if** $gps_{speedknown}$ **then**
8:            $v_t = N(v_{gps}, \sigma_v^2)$, where $\sigma_v = \frac{1}{5} \cdot \sqrt{\Delta_t}$
9:         **else**
10:            $v_t = N(v_{t-1}, \sigma_v^2)$, where $\sigma_v = 1 \cdot \sqrt{\Delta_t}$
11:         **end if**
12:      **end if**
13:    **else**
14:      $v_t = 0$
15:    **end if**
16:    with $\begin{cases} v_t = |v_t|, \text{ if } v_t < 0 \\ v_t = 2 \cdot v_{max} - v_t, \text{ if } v_t > v_{max} \end{cases}$
17:    **Bearing**
18:    **if** $gps_{bearingknown}$ **then**
19:      $\alpha_t = N(\sqrt{\alpha_{gps} \cdot \alpha_{t-1}}, \sigma_\alpha^2)$, where $\sigma_\alpha = 0.5\pi - \frac{\arctan\left(\frac{\sqrt{v_{gps}}}{2}\right)}{\sqrt{\Delta_t}}$
20:    **else**
21:      $\alpha_t = N(\alpha_{t-1}, \sigma_\alpha^2)$, where $\sigma_\alpha = 0.5\pi - \frac{\arctan\left(\frac{\sqrt{v_t}}{2}\right)}{\sqrt{\Delta_t}}$
22:    **end if**
23:    **Position**
24:    $\sigma_p = 0.5m$
25:    $\mathbf{X}_t = \begin{bmatrix} x_t \\ y_t \end{bmatrix} = \begin{bmatrix} x_{t-1} + v_t \cdot \cos(\beta_t) \cdot \Delta_t + n \\ y_{t-1} + v_t \cdot \sin(\beta_t) \cdot \Delta_t + n \end{bmatrix}$, where $n = N(0, \sigma_p)$
26:    **Floor Change**
27:    **if** ParticleInTransitionArea **then**
28:      Change z_t
29:    **end if**
30:    **Map Filtering**
31:    *valid* =checkIfParticleIsCrossingAWall
32:    **if** valid **then**
33:      return $\mathbf{x}_t^i$
34:    **end if**
35: **end for**
36: return $\mathbf{x}_t^i$

4.3.3. GSM

As described by [3], most GSM based positioning methods use the position of the different cell towers. This can be used in network-based or terminal-based positioning. For network-

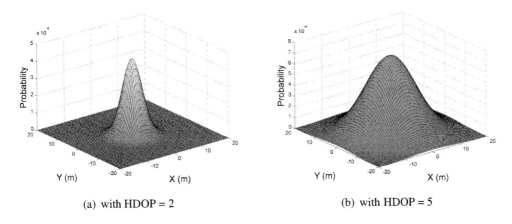

Figure 7. Example of GNSS measurement probability with different HDOP.

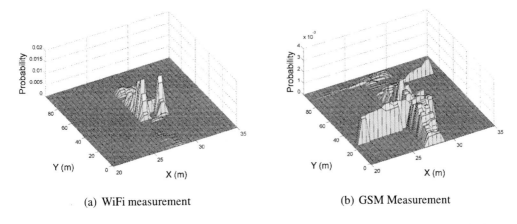

Figure 8. Example of WiFi and GSM measurement probability using fingerprinting.

based positioning, we need the cooperation of the mobile operator. This does not cope with the opportunistic concept of OLS where we want to grasp data using the devices. Due to the fact that we want to use standard hardware without requiring extra additional hardware, or information like the cell tower locations, from the mobile operator; TOA, TDOA, AOA and pure RSS based methods are not feasible either. The only viable method is to use a fingerprint based method similar to WiFi. The fingerprint can be used for Cell-ID localization or an RSS fingerprinting localization. Cell-ID localization will only identify the areas known as the cell used by the mobile device. Building a likelihood function using the fingerprint similar to WiFi, will result in a probability map as shown in Figure 8(b). The likelihood function of GSM, which is the same as WiFi [31], cannot identify such a specific area as WiFi does, which will result in less accurate localization.

4.3.4. Bluetooth

Bluetooth can be used as cell based localization where the location of static Bluetooth devices are known. A measurement model to implement this is described in [37]. A sample measurement probability can be seen in Figure 9. In Figure 9(a) the likelihood function, when a device at position (0,0) is discovered, is shown. A Class 2 Bluetooth device can be discovered up to 10 m distance in line-of-sight. A Sigmoid function is used to create a soft threshold between discoverable and non-discoverable distance (5). There is wall at $y = -3$. Since a wall attenuates the Bluetooth signal, the maximum discoverable distance will be lowered to 5 m if passing a wall (6). In Figure 9(b), two devices are discovered, one at (0,0) and one at (10,0). In the case of multiple devices the Likelihood Observation Function (LOF) for each device is multiplied to get an lof, and incorporates all discoverable devices.

$$y = \frac{1}{1+e^{x-10}} \quad (5)$$

$$y = \frac{1}{1+e^{x-5}} \quad (6)$$

Bluetooth can also be used to implement dynamic object binding [37]. In dynamic object binding the likelihood of a device is based on a likelihood of the bound object. For example, device A (e.g., a PDA) is located using its WiFi sensor, but also has a Bluetooth connection. Device B (e.g., a mobile phone) only has a Bluetooth connection. If device A can discover device B through Bluetooth, OLS can estimate the position using the likelihood function of device A. In this case, we can use the Bluetooth measurement model to estimate a likelihood of device B based on the likelihood of device A, which is estimated using the WiFi measurement model. An example of such a likelihood is shown in Figure 10.

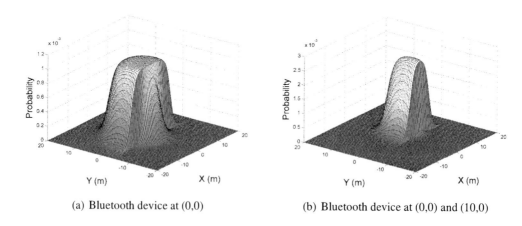

(a) Bluetooth device at (0,0)

(b) Bluetooth device at (0,0) and (10,0)

Figure 9. Example of Bluetooth measurement probability with a wall at y = -3.

4.3.5. Fusion of Measurement Models

Every measurement model produces a likelihood function. This likelihood observation function defines the probability of the device at every location in the environment. Fusing multiple measurements can be performed by multiplying the different likelihood observation functions. An example of a fused WiFi and GSM measurement probability is shown in Figure 11. The two bottom graphs show the individual LOFs for the WiFi and GSM measurement. The top graphs show the fused likelihood.

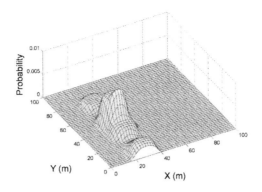

Figure 10. Example of Bluetooth measurement probability using object binding.

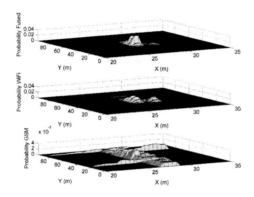

Figure 11. Example of fused WiFi and GSM measurement probability.

4.4. Overall Fusion Flow

Figure 12 shows an example on how data is handled when it arrives at the fusion engine. From left to right, the data (arriving at different time stamps) are shown. The way this data is handled, according to the different steps of a particle filter, is described next:

1. Knowledge about moving or standing still is saved to be used during later decisions.

2. Data from GPS can be used for the motion model (since we have an idea of the approximate speed and bearing), and afterwards in the GPS measurement model. After the measurement model, the particles are resampled.

3. If the PDR tells the engine that there was movement for a certain time, but is unable to estimate the distance traveled, the standard motion model will be used.

4. The WiFi data can be used directly in the WiFi measurement model, which is followed by the resampling.

5. If we get the estimated distance from the PDR, this is used in the PDR motion model.

6. If WiFi and GSM data are arriving at the same time interval, the measurement models is used directly after each other, which results in multiplication of both WiFi and GSM likelihood observation functions. After processing both, the resampling will select the samples with the best probability.

7. If PDR and GPS data is arriving together, the PDR motion model will use the GPS bearing as well. After which the GPS measurement model and the resampling is done.

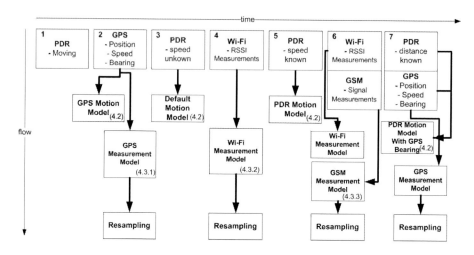

Figure 12. An example of the fusion flow.

5. Fingerprint Predictions for an Opportunistic Localization System

The opportunistic localization concept used in OLS tries to minimize the human involvement in setting and maintenance of the localization system. This is possible with many localization sensors currently used in OLS but obvious difficulties arise with WiFi and GSM localization, which need to have a signal fingerprint available for the estimation of a smart phone position. A possible solution is the use of predicted fingerprints rather than using the manually collected, where the prediction can be performed automatically by OLS based on environment description available to OLS system without any user involvement. Depending on the level of details of the environment description, the three different signal propagation prediction models are used (the One Slope Model, Multi-wall Model and Motif Model) leading to fingerprints of different levels of fidelity. It is therefore the objective of this chapter to compare the influence of predicted fingerprint fidelity on the WiFi based localization accuracy and to compare it also against a manually calibrated fingerprint.

5.1. Fingerprint Prediction

Indoor propagation is known as one of the most complicated propagation due to the various types of the building structures and materials used. Several indoor propagation models are implemented to predict the RSS fingerprint.

There are two conventional approaches to the modeling: empirical and deterministic. The advantages of empirical approach are speed, only simple input is needed and the formulas are simple to apply. The main disadvantages are poor site-specific accuracy and the incapability to predict wide band parameters of communication channel.

The deterministic models try to follow physical principles of electromagnetic wave propagation. The most popular are ray tracing and ray launching. They are based on geometrical optic principles. These models are very accurate, site-specific and can predict wide band parameters as a wall. On the other hand they are slow, the speed can be increased but some pre-processing and simplifications are necessary [47].

Furthermore, semi-deterministic models emerge to combine the empirical and deterministic approaches and gain their advantages. Introduced in [48], the Motif Model is a semi-deterministic model that is based on a ray launching technique, the Monte Carlo method and general statistics.

Utilizing the empirical model (i.e., one sloop model and multi-wall model) and the semi-deterministic model (i.e., motif model), this paper investigates the RSS prediction and its influence on the indoor location accuracy.

5.1.1. One-Slope Model

The One-Slope Model (1SM) [49] is the simplest way to compute the average signal level within a building without detailed knowledge of the building layout. The path loss is just a function of distance between transmitter and receiver antennas:

$$L(d) = L_0 + 10n\log(d) \qquad (7)$$

where L_0 (dB) is a reference loss value for the distance of 1 m, n is a power decay factor (path loss exponent) defining slope, and d (m) is a distance. L_0 and n are empirical parameters for a given environment, and fully control the prediction.

5.1.2. Multi-Wall Model

The Multi-Wall Model (MWM) [49] takes into account wall and floor penetration loss factors in addition to the free space loss (8). The transmission loss factors of the walls or floors traversed by the straight-line joining the two antennas are summed to give the total penetration loss (9) or (10), respectively. The signal loss is given by:

$$L_{MWM} = L_1 + 20\log(d) + L_{Walls} + L_{Floors} \qquad (8)$$

where

$$L_{Walls} = \sum_{i=1}^{I} a_{w_i} k_{w_i} \qquad (9)$$

$$L_{Floors} = a_f k_f \qquad (10)$$

L_{MWM} denotes the predicted signal loss (dB); L_1 is the free space loss at a distance of 1 m from the transmitter (dB); L_{Walls} is the contribution of walls to the total signal loss (dB); L_{Floors} is the contribution of floors to the total signal loss (dB); a_{w_i} is the transmission loss factor of one wall of ith kind (dB); k_{w_i} is the number of walls of its kind; I is the number of wall kinds; a_f is the transmission loss factor of one floor (dB); k_f is the number of floors.

Figure 13 shows a visualization of the MWM predicted fingerprint from 5 access points.

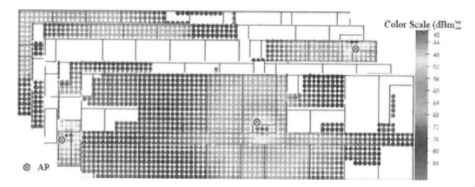

Figure 13. Predicted fingerprint with Multi-wall model.

5.1.3. Motif Model

The Motif Model (MM) is a combination of advantages of empirical and deterministic approaches. The model is based on a modified ray-launching technique. Ray propagation takes advantage of the modified simple line-drawing technique that divides the environment into a grid.

The fundamentals of the motif model algorithm are shown in Figure 14. A simple bitmap of the floor plan serves as the main input data. Filled pixels represent walls, partitions and obstacles. Different colors distinguish different materials. Propagation prediction is calculated in all empty elements at once.

The signal strength is predicted in each empty element of the bitmap by recording the number of passed rays. Predicted fingerprint with Motif model from 5 access points is visualized in Figure 15.

6. Experiments and Discussion

6.1. Opportunistic Data Fusion

Figure 16 shows a part of the Cork Institute of Technology (CIT) buildings where test runs where done. Every floor was 3.65 m high, the corridors were 2.45 m width and the part of the floors used were 44 meters wide and 110 m long. A 'quick-and-dirty' fingerprint was made for WiFi and GSM, in the corridor and the rooms that where accessible. The fingerprinting was done 2 months before the actual test measurement as described in this paper, to include the influence of signal changes due to environmental changes. The WiFi access points detected during fingerprinting and later in the experiments were a part of CIT's campus wide WiFi network maintained solely for the communication purpose and no considerations were made towards localization suitability during its design. There were, therefore, detected up to 3 but mostly only 2 AP at every point of the fingerprint. The

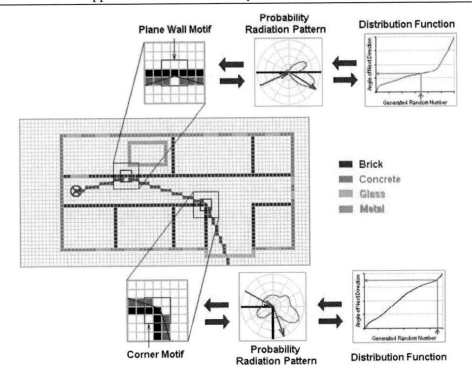

Figure 14. Fundamental of Motif algorithm.

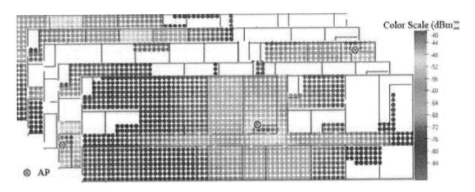

Figure 15. Predicted fingerprint with Motif model.

placement of the access points are shown in Figure 13 with a triangle. If the localization accuracy of WiFi should be improved, extra access points could be added, taking into account the geometric diversity of the access point locations. The locations of the GSM cell towers were unknown, but their position are not necessary for the localization.

The fingerprinting of WiFi and GSM was done while holding the Neo PDA, to be able to insert reference points by touching on the screen the current place on a detailed map. This fingerprinting could be improved by taking more measurements to collect more data to incorporate, for example, all different antenna orientations of the PDA. The main goal is to test the opportunistic localization using 'non-ideal data.' The localization experiments was done then with another Neo PDA having the PDA in a pocket of the trousers, which

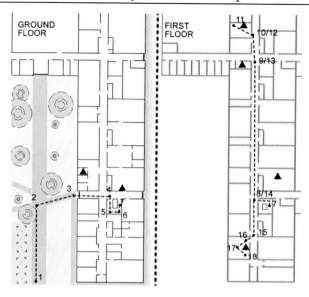

Figure 16. The part of 2 floors of the CIT building with the track of the test run, described in this paper.

influences the signal strength heavily, again to stress the 'non-ideal,' opportunistic concept.

Different test runs with other trajectories where made and all had comparable results. The trajectory discussed in this paper can be seen in Figure 16. This trajectory is chosen because it demonstrates the outdoor-indoor seamless transitions and the transition between floors. The test person came from outdoor, entered the corridor on the ground floor, climbed the stairs to the first floor, walked along the corridor to an office, turned there and walked back along the same corridor to another office. The person was constantly in motion, which makes localization estimation by the fusion engine harder.

Figure 17 and 19 and Table 3 give an overview of the mean estimation error, split up in the different areas (outdoor, first floor, and second floor). All systems use map filtering. As expected it is the fusion of all possible information (*WiFi+GSM+GPS+PDR*) which gives the best results. The improvement using PDR towards the fusion without PDR (*WiFi+GSM+GPS*) is mostly visible in the indoor part, since in outdoor, the speed and bearing of GPS is used in the motion model.

WiFi performs best inside, while GPS only works outside and in the first few meters behind the entrance door. Since the Neo PDA does not contain an ultra-sensitive GPS receiver and the GPS signals are not able to penetrate the roof and walls enough to be received by the PDA, GPS cannot be used on any other indoor location in this test. Furthermore the PDA is located in the trousers pocket of the test person, which makes it even harder to receive any GPS signals. GSM localization using fingerprinting (pattern matching) will work best indoors, but is most powerful in combination with WiFi since their likelihood observation functions (LOFs) often complement well each other.

Figure 18 shows two consecutive measurements, the left one a PDR+GSM measurement and the right one PDR+WiFi. The cross depicts the real position, the black dot the estimated position. The dark grey cloud are the particles, the light grey dots represent the

OLS - Opportunistic Localization System for Smart Phones Devices 215

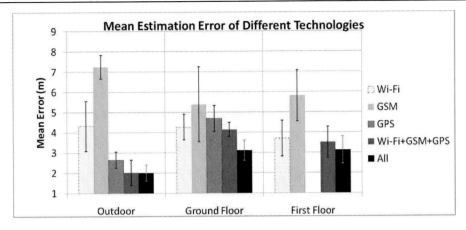

Figure 17. Results of location estimation using different technologies.

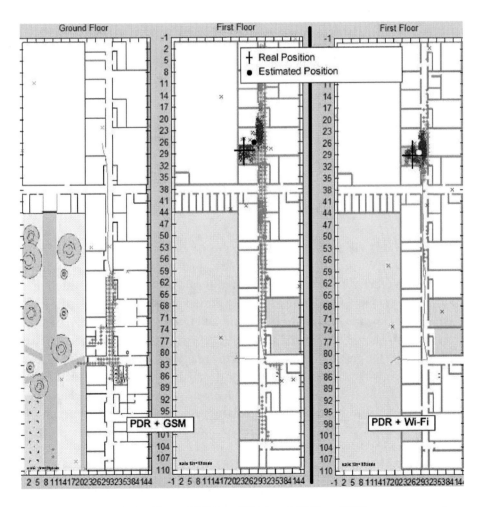

Figure 18. Particle Filter of PDR + GSM and PDR + WiFi measurement.

likelihood observation function (LOF) of the measurement (more dense = more likely). The ground floor is not shown in the WiFi figure since the LOF is equal to zero for all places on the ground floor.

The particles distribution at the left side (PDR+GSM) is the result of the position of the particles in the previous particle set, the motion model defined by PDR (moving = true, speed = 1.7 m / 1.47 s) and the measurement model defined by the GSM measurement (shown in Table 1) and the resampling step.

The particle distribution at the right side is formed starting from the particle distribution after the prevoius step described, the motion model defined by PDR (moving = true, speed = 1.6 m / 1.37 s) and the measurement model defined by the WiFi measurement (shown in Table 2) and the resampling step.

Table 1. Example GSM Measurement

Cell Tower ID	Field Strength
5-37231-789	-90 dBm
24-25025-17	-100 dBm
3-27642-799	-106 dBm
3-2764-20	-100 dBm

Table 2. Example WiFi Measurement

WiFi Access Point	Signal Strength
00:20:A6:62:87:1B	-61 dBm
00:20:A6:63:1D:4F	-93 dBm

As shown in 3, it is visible that GSM has more problems detecting on which floor the object is located thenWiFi. This is because the signal strengths of GSM on two floors directly under each other, does not differ very significantly if the floor construction is comparable with each other and GSM signal is likely horizontally propagating from the side of the building. Again the combination of both WiFi and GSM gives an improvement.

The results also show that the addition of PDR give a slight improvement since the motion model guides the particles better towards the real position of the object; for example, around the staircase where the particles can go from the ground to the first floor. The occurrence of estimating the wrong floor, only happens around the time where the object moves from one floor to the other. During normal movement of the object, the motion model does not allow switching floors if the particles are not in the neighborhood of a transition area (stairs or elevator).

The accuracy of the WiFi and therefore also of the fusion using WiFi together with another technology, can be greatly improved by adding extra access points, since in most of the locations only 2 access points are visible. The accuracy improves to about a meter if the person is standing still for a while, which gives the particles the opportunity to converge around the real location.

Table 3. Results of location estimation using different technologies. (in meter)

Technology	Mean Error	Std. Dev.	Outdoor	Floor 0	Floor 1	Correct Floor
GPS	3.08	1.31	2.65	4.69	N/A	N/A
GSM	6.18	2.36	7.24	5.39	5.82	70%
WiFi	3.90	1.88	4.32	4.27	3.70	88%
WiFi+GSM+GPS	3.05	1.49	2.02	4.14	3.51	91%
WiFi+GSM+GPS+PDR	2.73	1.28	2.01	3.10	3.12	93%

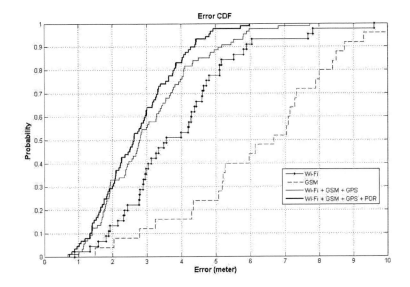

Figure 19. CDF Error of location estimation using different technologies.

6.2. Fingerprint Prediction

The fingerprint prediction fidelity influence was evaluated in a smaller scale experiment covering the first floor used in the overall experiment. The RSS fingerprint was mapped on the floor-plan divided into the 1 m square uniform-grid. Figure 20 illustrated measured fingerprint in the first floor floor-plan.

RSS fingerprint prediction is performed with One-Slope model, Multi-Wall model, and Motif Model. The gauges of fingerprint accuracy are the mean μ and standard deviation σ of the difference between the signal measurements and the prediction. Fingerprint prediction accuracy is summarized in Table 4.

Table 4. Fingerprint accuracy

	1SM	MWM	Motif Model
Difference (dBm)	$\mu = -7.45$	$\mu = -5.53$	$\mu = -1.76$
	$\sigma = 7.24$	$\sigma = 7.12$	$\sigma = 6.45$

Figure 20. Measured-fingerprint in the Electronic Department Building.

Furthermore the predicted and the measured fingerprint are used in the location system to estimated user location. Prerecorded real measurement data is reused during off-line tracking to evaluate location accuracy for each RSS fingerprints and location techniques. The location accuracy is summarized in Table 5.

Table 5. Location accuracy (meter)

	Measured	Motif Model	MWM	1SM
Particle Filter	$\mu = 1.98$	$\mu = 3.18$	$\mu = 6.40$	$\mu = 13.70$
	$\sigma = 1.39$	$\sigma = 2.08$	$\sigma = 9.08$	$\sigma = 10.90$

Figure 21 illustrated the CDF of location accuracy with Particle Filter algorithm for measured and predicted fingerprints.

One-slope model produces the poorest indoor location performance. The accuracy is improved in the Multi-wall model. Semi-deterministic Motif model has the best accuracy with the Particle Filter algorithm ($\mu = 3.18$ and $\sigma = 2.08$) among the predicted fingerprints. Furthermore, the influence of fingerprint prediction fidelity on the location accuracy is investigated. The fingerprint mean error μ of 0 (summarized in Table 6) is acquired to achieve that purpose.

Table 6. Fingerprint accuracy

	1SM	MWM	Motif Model
Difference (dBm)	$\mu = 0$	$\mu = 0$	$\mu = 0$
	$\sigma = 7.24$	$\sigma = 7.12$	$\sigma = 6.45$

The location accuracy with fingerprint mean error μ of 0 is summarized in the following Table 7. As it is seen in the Table 7, the location accuracy is improved for 1SM, MWM, and Motif Model. It shows that better fingerprint accuracy will lead to better location prediction. The following figures depict the trajectory and error vector of the predicted location in the Electronic Engineering building. A walk around building is performed from a room and a corridor. The vector origin shows the real user trajectory and the vector end point is the

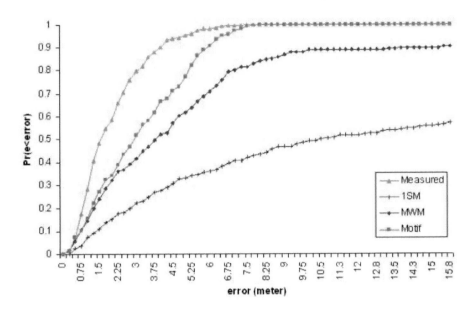

Figure 21. Location accuracy with Particle Filter Algorithm.

Table 7. Fingerprint accuracy

	1SM	MWM	Motif Model
Particle Filter	$\mu = 3.47$	$\mu = 3.03$	$\mu = 2.94$
	$\sigma = 2.45$	$\sigma = 2.01$	$\sigma = 1.89$

predicted location. Figure 22 shows the user trajectory with measured fingerprint.

Figure 23 shows the user trajectory with 1SM predicted fingerprint PF algorithm.

Figure 24 shows the user trajectory of with MWM predicted fingerprint.

Figure 25 illustrates the user trajectory with MM predicted fingerprint.

Despite that the measured-fingerprint still has the overall best accuracy ($\mu = 1.98$ and $\sigma = 1.39$), semi-deterministic prediction with Particle Filter technique seems to offer an appealing method in deploying indoor location. It is due to the speed and flexibility of the calibration phase without compromising accuracy. The predicted fingerprint is highly adaptive to any major change in propagation environment. Environment changes (e.g., wall addition or access point relocation) can be accommodated by adjusting the floor-plan drawing and launching the signal prediction afterwards.

7. Conclusion

This chapter demonstrated the concept of opportunistic localization for smart phones devices where the location estimation is done only with the available opportunistic data, depending on the environment and the client's hardware capabilities. A particle filter with an

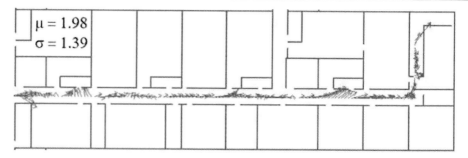

Figure 22. Trajectory and error vector with the measured fingerprint.

Figure 23. Trajectory and error vector with the 1SM predicted fingerprint.

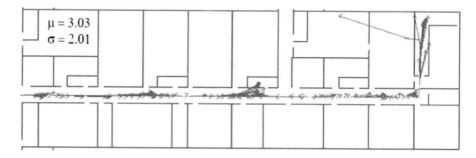

Figure 24. Trajectory and error vector with the MWM predicted fingerprint.

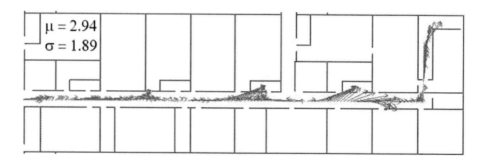

Figure 25. Trajectory and error vector with the Motif model predicted fingerprint.

adaptive motion and measurement model is used to fuse the different sensor data. We have shown the resulting localization when a mobile device uses a separate technology, and the results, if a device is able to collect more data so the fusion engine can combine them.

We have focused on the opportunistic concept, where the sensor data measurement, fingerprinting and used hardware should represent realistic non-ideal data. We did not focus on acquiring a very high accuracy, since this can be obtained by installing for example extra access points specially deployed for improving the WiFi coverage for localization. Where the user is not interested in installing any extra hardware to be able to have a localization service, but where the service will adapt to the environmental opportunities. The tests in Cork Institute of Technology, Ireland, show localization using GSM with a mean error of 6.18 m using standard known signals of up to 8 surrounding cell towers, for WiFi around 3.90 m where in most cases only 2 access points where visible. The fusion of GSM, WiFi, GPS, PDR and map filtering, gives an improvement, which results in a mean error of 2.73 m and a correct floor detection of 93%. Most applications show a benefit from opportunistic localization where the installation of dedicated localization hardware is not needed to improve the localization accuracy, and a mean error around 3 m will allow sufficient data to build application using the opportunistic data.

Moreover, our work in investigating the influence of measured and predicted fingerprint is presented. Several propagation models and their performance in OLS are highlighted. It is found that better fingerprint prediction fidelity can lead to not only improved accuracy, but also a scalable and adaptive location system therefore suitable to the opportunistic concept presented in this chapter. It is shown that the semi-deterministic model like the Motif Model has better accuracy and flexibility compared to other models. It is therefore more suitable for fingerprint prediction assuming knowledge of main obstacles layout in the area of fingerprint prediction. It is not a hard assumption as environment layouts and maps are usually required for the location estimate presentation to the end user in the context of environment, as well as for constraining the object motion as presented earlier.

References

[1] S. Beauregard, "Omnidirectional pedestrian navigation for first responders," in *Positioning, Navigation and Communication, 2007. WPNC'07. 4th Workshop on*, 2007, pp. 33–36.

[2] M. Porretta, P. Nepa, G. Manara, and F. Giannetti, "Location, location, location," *IEEE Vehicular Technology Magazine*, vol. 3, Issue 2, pp. 20–29, 2008.

[3] N. Deblauwe, "Gsm-based positioning: Techniques and application," Ph.D. dissertation, VUB, 2008.

[4] Widyawan, M. Klepal, and S. Beauregard, "A novel backtracking particle filter for pattern matching indoor localization," in *MELT '08: Proceedings of the first ACM international workshop on Mobile entity localization and tracking in GPS-less environments*. New York, NY, USA: ACM, 2008, pp. 79–84.

[5] "Ekahau," August 2010. [Online]. Available: http://www.ekahau.com/

[6] "Airoscout," August 2010. [Online]. Available: http://www.airoscout.com

[7] A. LaMarca, Y. Chawathe, S. Consolvo, J. Hightower, I. Smith, J. Scott, T. Sohn, J. Howard, J. Hughes, F. Potter et al., "Place lab: Device positioning using radio beacons in the wild," in *Proceedings of PERVASIVE*, vol. 3468, 2005, pp. 116–133.

[8] "Skyhook wireless," August 2010. [Online]. Available: http://www.skyhookwireless.com

[9] F. Alizadeh, "Skyhook: Opportunistic and Hybrid Localization," Presented at Invitational Workshop on Opportunistic RF Localization for Next Generation Wireless Devices, June 2008.

[10] T. Young, "Rosum: Tv+gps location and timing," Presented at Invitational Workshop on Opportunistic RF Localization for Next Generation Wireless Devices, June 2008.

[11] "Rosum," August 2010. [Online]. Available: http://www.rosum.com/

[12] D. Hall and J. Llinas, "An introduction to multisensor data fusion," *Proceedings of the IEEE*, vol. 85, no. 1, pp. 6–23, 1997.

[13] D. L. Hall and J. Llinas, *Handbook on Multisensor Data Fusion*. CRC, 2001.

[14] D. L. Hall and T. S. Shaw, "Multi-Sensory, Multi-Modal Concepts for Information Understanding," in *RTO-MP-105, Proceedings of RTO IST Workshop on "Massive Military Data Fusion and Visualisation: Users Talk with Developers"*. Halden, Norway: NATO, September 2002.

[15] P. Moutarlier and R. Chatila, "Stochastic multisensory data fusion for mobile robot location and environment modeling," in *5th Int. Symposium on Robotics Research*, vol. 1, 1989.

[16] J. Castellanos, J. Neira, and J. Tards, "Multisensor fusion for simultaneous localization and map building," *IEEE Transactions on Robotics and Automation*, vol. 17, pp. 908–914, 2001.

[17] J. Hightower and G. Borriello, "Particle filters for location estimation in ubiquitous computing: A case study," in *Proceedings of the 6th Ubicomp*, September 2004.

[18] F. Zorzi and A. Zanella, "Opportunistic localization: modeling and analysis," in *Proc. of the IEEE 69th Vehicular Technology Conference (VTC Spring 2009)*, 2009.

[19] F. Zorzi, G. Kang, T. Prennou, and A. Zanella, "Opportunistic Localization Scheme Based on Linear Matrix Inequality," in *IEEE International Symposium on Intelligent Signal Processing, 2009 - WISP 2009*, Budapest, Hungary, August 2009.

[20] G. Kang, T. Prennou, and M. Diaz, "An opportunistic indoors positioning scheme based on estimated positions," in *Proceedings of the IEEE Symposium on Computers and Communications - ISCC'09*, Sousse, Tunisia, July 2009.

[21] E. Duff, J. Roberts, and P. Corke, "Automation of an underground mining vehicle using reactive navigation and opportunistic localization," in *Proc. 2002 Australasian Conference on Robotics and Automation*, Auckland, Australia, November 2002.

[22] H. Lee and H. Aghajan, "Collaborative node localization in surveillance networks using opportunistic target observations," in *Proceedings of the 4th ACM international workshop on Video surveillance and sensor networks.* ACM, 2006, p. 18.

[23] J. McIntosh, "Passive three dimensional track of non-cooperative targets through opportunistic use of global positioning system (GPS) and GLONASS signals," US Patent 6,232,922 B1, 2001.

[24] A. Ofstad, E. Nicholas, R. Szcodronski, and R. Choudhury, "Aampl: Accelerometer augmented mobile phone localization," in *Proceedings of the first ACM international workshop on Mobile entity localization and tracking in GPS-less environments.* ACM, 2008, pp. 13–18.

[25] S. Couronne, N. Hadaschik, M. Fassbinder, T. von der Grun, M. Klepal, Widyawan, M. Weyn, and T. Denis, "Locon: a platform for an inter-working of embedded localisation and communication systems," in *Proceeding of 6th Annnual IEEE SECON 2009*, Rome, Italy, June 2009.

[26] H. Millner, P. Gulden, M. Weyn, M. Fassbinder, and A. Casaca, "Locon deliverable d4.2: Description of the locon protocol," Symeo, CIT, CEA-LETI, INOV, Artesis, Fraunhofer IIS, Tech. Rep., 2009.

[27] F. van Diggelen and C. Abraham, "Indoor gps technology," *CTIA Wireless-Agenda, Dallas,*, May 2001.

[28] M. Weyn and F. Schrooyen, "A wifi-assisted-gps positioning concept," in *Proceeding of ECUMICT*, Gent, Beglium, March 2008.

[29] S. Wibowo, M. Klepal, and D. Pesch, "Time of flight ranging using off-the-shelf wifi tag," in *Proceedings of PoCA 09*, 2009.

[30] Widyawan, M. Klepal, and D. Pesch, "A bayesian approach for rf-based indoor localisation," in *IEEE ISWCS*, Trondheim, Norway, October 2007.

[31] Widyawan, "Learning data fusion for indoor localisation," Ph.D. dissertation, Cork Institute of Technology, 2009.

[32] T. Wigren, "Adaptive enhanced cell-id fingerprinting localization by clustering of precise position measurements," *IEEE Transactions on Vehicular Technology*, vol. 56, pp. 3199–3209, 2007.

[33] J. Caffery and G. Stuber, "Overview of radiolocation in cdma cellular systems," *IEEE Communications Magazine*, vol. 36, no. 4, pp. 38–45, 1998.

[34] A. Varshavsky, E. de Lara, J. Hightower, A. LaMarca, and V. Otsason, "Gsm indoor localization," *Pervasive and Mobile Computing*, vol. 3, no. 6, pp. 698–720, 2007.

[35] I. Bylemans, M. Weyn, and M. Klepal, "Mobile phone-based displacement estimation for opportunistic localisation systems," in *Proceedings of UBICOMM9: International Conference on Mobile Ubiquitous Computing, Systems, Services and Technologie*, Malta, October 2009.

[36] J. Hallberg, M. Nilsson, and K. Synnes, "Positioning with bluetooth," in *10th International Conference on Telecommunications. ICT 2003*, 2003, pp. 954–958.

[37] I. De Cock, W. Loockx, M. Klepal, and M. Weyn, "Dynamic object binding for opportunistic localisation," in *Proceedings of UBICOMM2010: International Conference on Mobile Ubiquitous Computing, Systems, Services and Technologie*, Florence, Italy, October 2010.

[38] C. Settgast, M. Klepal, and M. Weyn, "Object Correlation Evaluation for Location Data Fusion," in *The International Conference on Ubiquitous Positioning, Indoor Navigation and Location-Based Service, UPINLBS 2010*, Finland, October 2010.

[39] S. Thrun, W. Burgard, and D. Fox, *Probabilistic Robotics (Intelligent Robotics and Autonomous Agents)*. The MIT Press, 2006.

[40] D. Fox, "KLD-sampling: Adaptive particle filters and mobile robot localization," *Advances in Neural Information Processing Systems (NIPS)*, pp. 26–32, 2001.

[41] A. Soto, "Self adaptive particle filter," in *International Joint Conference on Artificial Intelligence*, vol. 19. Citeseer, 2005, p. 1398.

[42] M. Weyn, M. Klepal, and W. Widyawan, "Adaptive motion model for a smart phone based opportunistic localization system," in *MELT'09: Proceedings of the 2nd international conference on Mobile entity localization and tracking in GPS-less environments*. Berlin, Heidelberg: Springer-Verlag, 2009, pp. 50–65.

[43] B. Ristic and S. Arulampalam, *Beyond the Kalman filter: Particle filters for tracking applications*. Artech House, 2004.

[44] N. Gordon, D. Salmond, and A. Smith, "Novel approach to nonlinear/non-Gaussian Bayesian state estimation," in *IEE Proceedings*, vol. 140, no. 2, 1993, pp. 107–113.

[45] P. Fearnhead, "Sequential monte carlo methods in filter theory," Ph.D. dissertation, Mercer College, University of Oxford, 1998.

[46] D. L. Wilson, "HDOP and GPS Horzintal Position Errors," August 2010.

[47] P. Pechac and M. Klepal, "Effective indoor propagation predictions," in *IEEE VEH TECHNOL CONF*, vol. 3, no. 54, 2001, pp. 1247–1250.

[48] M. Klepal, "Novel approach to indoor electromagnetic wave propagation modelling," Ph.D. dissertation, Czech Technical University, Department of Electromagnetic Field, Prague, Czech, 2003.

[49] "Cost231 final report, digital mobile radio: Cost231 view on the evolution towards 3rd generation systems," European Commission / COST Telecommunications, Brussels, Tech. Rep., 1998.

Chapter 8

MOBILE PHONES AND INAPPROPRIATE CONTENT

J. G. Phillips[1], P. Ostojic[1] and A. Blaszczynski[2]

[1]School of Psychology and Psychiatry, Monash University, VIC, Australia
[2]School of Psychology, University of Sydney, NSW, Australia

ABSTRACT

The mobile phone is touted to bridge the digital divide, with mobile phones being possessed by most individuals regardless of their age or ability. Improvements in phone and network capability have meant that most web content is now accessible anywhere by anyone at anytime. The degree to which the digital divide is being bridged has caused problems for legislators seeking to control access to specific content or activities (e.g. gambling; pornography) by sections of the community, particularly those most vulnerable (i.e. children and adolescents). The process of access to inappropriate content, or the excessive participation in activities, actually requires the transmission of a number of messages. This chapter discusses methods of controlling access to inappropriate content and restricting excessive activities. Providers of inappropriate content can be blocked at the source. Regulators have had success prosecuting providers of specific content (e.g. missed call scams), but jamming systems lack specificity. The use of contracts has tended to limit access to users of legal age, but may be circumvented by the use of debit cards, disposable and prepaid phones. Biometric systems have been under consideration to determine age, but are not foolproof. Attempts to control inappropriate content are compromised by providers operating off-shore.

MOBILE PHONES AND INAPPROPRIATE CONTENT

Mobile phone technology is likely to address concerns that a section of the community may not have access to the benefits of digital technology. Concerns that a section of the community might be technologically disenfranchised are countered by observations that in both Western and less developed nations more people have access to mobile phones than people accessing the internet via conventional computers. Hence it is the mobile phone that may bridge the digital divide (Boyera, 2006). The increasingly pervasive democratisation of this technology potentially allows anyone to access the internet anywhere and at any time. However the increased access challenges jurisdictions' ability to control the access to specific content (e.g. pornography, gambling) by specific groups (i.e. underage individuals). The present paper discusses methods of restricting access to inappropriate content.

NETWORK CAPABILITY

Internet access can be achieved through a number of technologies including dial-up on a copper wire network, wireless and optical fibre cable. The difference between technologies primarily relates to the speed at which data can be transmitted. The more data to be transmitted the faster the internet speed needs to be in order to achieve an acceptable user experience. Video and audio streaming of a live event in real time (e.g. a horse race) involves large amounts of data and hence requires a very fast "broadband" technology such as Asymmetrical Digital Subscriber Line (ADSL), optical fibre cable or satellite.

Definitions of exactly how fast an internet connection needs to be in order to qualify as "broadband" varies from country to country and changes as technology advances. In 2003 for example, the International Telecommunication Union's Standardization Sector defined broadband as a "transmission capacity that is faster than….1.5 or 2.0 Megabits per second (Mbits)" (ITU, 2003) whereas in 2009 the United States Federal Communications Commission defined "broadband" at the much slower rate of

"…transmission speeds exceeding 200 kilobits per second (Kbps)" (US-FCC, 2009).

There are a variety of methods of quantifying internet connectivity (http://en.wikipedia.org/ wiki/ Internet_access_worldwide Accessed: 20/10/10), but developed countries have greater access to the internet than undeveloped countries (http://www.nationmaster.com/graph/int_int_int_ban_mbp_percap-international-bandwidth-mbps-per-capita   Accessed: 20/10/10). *More people own mobile phones (67.6% world wide*, http://en.wikipedia.org/wiki/List_of_countries_by_number_of_mobile_phones_in_use Accessed: 20/10/10) *than people accessing the internet via conventional computers (28.7% worldwide* http://en.wikipedia.org/wiki/List_of_countries_by_number_of_Internet_users Accessed: 20/10/10).

In Australia, "broadband" involves speeds ranging from 256 kilobits/sec (kbps) to 24 megabit/sec (mbps) and utilises technologies including Hybrid Fibre Coaxial (HFC) cable, ADSL, Wireless and Satellite (ACMA, 2008). It should be noted however that the Australian Federal Government released its optical fibre-based "super fast" National Broadband Network plan in April 2009. That network is to deliver internet speeds of up to 100 megabits/sec to 90% of the Australian population with the remainder, those living in remote parts of the country, able to access the internet at speeds of 12 megabits/sec through next generation wireless and satellite technologies. While the Australian Communications and Media Authority (ACMA) reported that by 30 June 2008 all Australians had access to "broadband" internet through, at a minimum, relatively slow satellite technology (ACMA, 2008), the National Broadband Network aims to ensure "Every person and business in Australia, no-matter where they are located, will have access to affordable, fast broadband at their fingertips" (Media Release, 2009). Currently however, some 72% of Australian households have home internet access and approximately 86% of those (i.e. an estimated 5 million households) access broadband internet (Australian Bureau of Statistics, 2008-09).

The original purpose of the internet was to allow communication between a number of computers in physically different locations and so the internet was initially only accessible via computer. Use of the internet then was restricted to those able to afford such exotic and expensive machines. Scientific and technical advances eventually led to the ability to mass-produce computer chips with the result that not only did computers became more affordable and commonplace, but computer technology began to be incorporated into other devices. As a consequence a number of devices other than computers are now capable of accessing the internet with the most pervasive of all being mobile devices such as mobile phones and Personal Digital Assistants (PDA's). The type of information available and the user experience associated with accessing that information on such small, hand-held devices evolved as both hardware and software advanced.

Early "first generation" communications platforms were not well suited to internet traffic since they utilised analogue technology. "Second generation" (2G) platforms involved digital technology and resulted in, amongst other advances, a Global System for Mobile (GSM) communications and the ability to transmit simple, low volume digital data such as SMS's. Telecommunications carriers now provide "Third generation" (3G) digital technology platforms capable of handling modern data-rich internet traffic. Such platforms generally use Wideband Code Division Multiple Access (W-CDMA) technology to deliver download speeds of up to 14.4Mbit/s and so are capable of quickly and efficiently delivering both voice and data to mobile users (ACMA, 2009a). However the simple delivery of sufficient data to a device does not guarantee the device is able to display that data in meaningful way.

In order for internet sites to be correctly displayed on the small screens of mobile devices having comparatively poor computational power, the Wireless Application Protocol (WAP) was developed. WAP is an international standard specifically designed to allow handheld, mobile devices such as mobile phones and PDA's access to certain internet sites. Those sites use text and graphics specifically created for the small screens and limited computing power of such devices to ensure the dimensions of the site are such that it displays correctly on those devices.

Hardware and software advances such as those, along with continuing improvements in micro-processor technology, have resulted in the current situation whereby mobile devices can not only access broadband internet from most locations, but also provide the user with a high-quality experience as well. Current broadband technology permits large amounts of data to be broadcast including the simultaneous transmission of high quality video and audio data in real time with the user experience likely to improve with further advances in technology. Ostensibly then people have the ability to access high quality, live, interactive, internet-based applications 24 hours a day, 7 days a week regardless of their geographical location through a variety of devices such as portable laptops, PDA's and mobile phones.

While some applications pose no legal or ethical problems (e.g. sports scores, weather updates) some applications have been used inappropriately to trick consumers into making "purchases" on their mobile phones (http://www.scamwatch.gov.au/content/ index. phtml/tag/MobilePhoneScams Accessed: 23/10/10). Unlike the internet (Kraut, Sunder, Telang, and Morris, 2005), messaging over mobile networks can incur appreciable costs (Mahatanankoon, Wen, and Lim, 2005), serving as a means of funds transfer (Griffiths, 2007). Some scammers use premium rate messaging services as a method of robbing consumers. In missed call scams a consumer's phone is rung quickly such that the phone registers a missed call. People are then tricked into calling the number back to find out who it is. The number that is called back may be redirected by scammers to a premium rate service incurring appreciable costs per minute. In variations of this scam consumers are effectively paying to receive marketing messages. Sometimes the scammers tell the consumers that they have won a prize. The number to be called to 'claim' the prize is a premium rate number again leading to appreciable costs per minute. Other applications such as Trivia competitions resemble gambling as prizes are offered for essentially chance outcomes. Consumers are encouraged to receive premium rate "entry blanks" to a competition involving a prize. Consumers are misled as to their chances of winning, or how much it will cost to take part (Griffiths, 2007). *Where these activities occur within a government's jurisdiction, prosecutions can occur and companies can be fined* (http://www.amta.org.au/articles/amta/ ACMA.moves.to.stop.missed.call.marketing Accessed: 21/10/10).

Other applications such as violent computer games, pornography or gambling are more problematic because there may be difficulties controlling who views this content. Although the size of mobile devices notionally makes viewing content on their small screens a "private" affair, their mobility and ubiquity potentially mean unacceptable content can be accessed in inappropriate public locations such as schools (Finn, 2005) and so result in unintended exposure to minors. Indeed access to inappropriate content causes concern for community groups, and in some cases can lead to licencing being compromised or revoked.

For instance, casino style gambling over the internet is restricted in a number of jurisdictions (e.g. USA, Australia) but the access afforded by the internet has created technological loopholes (Eadington, 1988; 2004; Griffiths, 2003; Parke and Griffiths, 2004; Watson, Liddell, Moore, and Eshee, 2004) that can circumvent existing controls. Indeed the actions of illegitimate operators causes problems for recognised companies that are seeking to establish an internet presence (Eadington, 2004; Scoolidge, 2006). In Australia expenditure on legal forms of gambling has been estimated at 19 billion dollars (Productivity Commission, 2010). However, expenditure on illegal forms of online gambling in Australia has been estimated at 790 million dollars (Productivity Commission, 2010). As other forms of gambling are legal in Australia, and are a source of revenue for the government, the illegal forms of online gambling currently constitute a loss of revenue.

AGE VERIFICATION ONLINE

Underage access is a potential concern for providers of certain services such as gambling, and is certainly an issue raised by lobbyists. As younger cohorts are perceived to be the more likely users of this technology, the concern is not trivial. Previously not enough had been done to control the access of minors to inappropriate content (Byron, 2008) and so this has become an area of considerable effort (ACMA, 2009b). The focus of that effort appears to be on age verification whereby owners of a site involving age-sensitive content attempt to verify the age of individuals wishing to access that content before access is allowed. A number of approaches aimed at verifying user age remotely and accurately over the internet have been tried but they are not without their legal and ethical concerns.

There are a variety of means whereby a user's age can be verified online. Users can verify that they are overage. Alternatively the user's age can be verified from financial transactions. Biometrics and filters are other systems that can be employed to verify age. Surveillance is also likely to reduce access to inappropriate content. Each approach will be discussed in turn.

USER SELF-VERIFICATION

The simplest way of establishing someone's age is simply to ask them. In the case of some restricted content however, establishing the age of intending users while important, is not sufficient. Interactive gambling operators for example not only need to establish whether would-be patrons are of legal age to gamble, but also whether they are located in a jurisdiction that allows remote gambling and are not using the gaming venue for a fraudulent or prohibited activity such as "laundering" illicit money in an effort to avoid leaving "...a recognisable audit trail" (Taylor, 2003).

Such issues are currently typically addressed by site operators in statements included in their "Terms and Conditions" (or "End User Statement", "Player agreement" etc.). Such statements typically make reference to the site only being available to people over the legal minimum age in the jurisdiction in which they live, that the laws of that jurisdiction make it legal to access the content and that there is no criminal intent associated with accessing the content (see for example, the "Terms and Conditions of Use" for Party Poker, an internet casino licensed in Gibraltar (https://secure.partyaccount.com/about/legal_information_s.do Accessed: 23/09/10).

Intending users are required to agree to the operators Terms and Conditions before being allowed access to the site. Agreement is given by users ticking a box labelled "I agree" (or something similar) while those opting for the "I do not agree" option are prevented from accessing the site. In the case of remote gamblers, access to the site may be allowed but they are prevented from downloading proprietary software and so cannot gamble at the venue.

Such self-verification and authentication relieves site owners of the need to access the veracity of the information supplied by potential visitors making them solely reliant on the honesty of the intending user. While many of the shortcomings of such an approach are clear, a less obvious outcome in relation to gambling sites is that minors can be taken advantage of. Some gambling sites require patrons to pay money into the casino account prior to gambling, with any winnings paid into the users' personal account. Under-age visitors claiming to be of legal gambling age then are *able to gamble but cannot collect their winnings* since, as minors, they may not be able to access a personal bank account.

In establishing the jurisdiction of a would-be user it should be noted that site operators are not restricted to the honesty of the intending user. When a user requests a URL on their mobile handset the request and identifying information about the user are forwarded to the internet service provider (ISP). The identifying information is used by the ISP to bill the customer and is kept within their domain to protect customer privacy. With identifying information removed the URL request is subsequently submitted to the internet along with an internet protocol (IP) "address". An IP address is a series of unique numbers used to identify the device making the request and is needed so that the requested site knows where to send the response.

Assigning IP addresses to countries is the task of the Internet Assigned Numbers Authority (IANA). For example, internet service providers in Australia have been assigned IP addresses within various numerical ranges including those from 58.65.248.0 to 58.65.255.255 (see http://www.ipaddresslocation.org/ip_ranges/get_ranges.php). Addresses within a range are further sub-divided and allocated to a specific physical site housing telecommunications hardware such as routers, servers, switches etc. Since the assigned number ranges are both country-specific and public knowledge it is a simple matter for off-shore service providers to determine the country of origin of a mobile-handset generated internet request. The service provider captures the IP address associated with the request and uses geo-location software to determine the origin of that request. The accuracy with which the origin of the request can be located depends on the software used. "IP2Location" for example appears to have established databases of the physical location of specific address ranges over a period of time and claims to be able to provide not only the country, state and city of origin but also latitude, longitude, postcode, time zone, connection speed, ISP and domain name, international direct dial country code (e.g. 61 for Australia), area code and both the name and identity number of the nearest weather station (see http://www.ip2location.com/)

While it is then technically possible for site operators to identify and block mobile-handset generated internet requests originating from *countries* where access to such content is prohibited (e.g. gambling), such action involves costs. Costs include not only the cost of buying and maintaining the capture/geo-location software but also the cost of reduced revenue resulting from taking such action. In addition, the difficulty in enforcing national laws internationally means there are few legal consequences to operators who provide access to sites in response to requests originating in countries where such access is prohibited by law. Indeed it has been argued (USA vs Antigua and Barbuda) that *prohibitions of online gambling constitute a restriction of trade* (http://www.wto.org/english/tratop_e/dispu_e/cases_e/ds285_e.htm).

FINANCIAL

Sites that are notionally restricted to adults typically require users to pay for much of the content offered on the site and so restricting access to minors may involve attempts to verify age through the use of credit cards (ACMA, 2009b, p. 43). Such systems can, however be easily circumvented by prepaid or debit cards, and are vulnerable to other problems. The use of a credit card to verify identity can lead to inappropriate charges being made. For instance a recent "market research" scam purporting to represent McDonalds asked respondents for credit card details to enable a $50 "payment", but instead led to inappropriate charges being made against victim's credit cards (http://www.acma.gov.au/WEB/STANDARD/pc=PC_310566). Some countries have developed national identity card systems, but uptake appears to be low (ACMA, 2009b, p. 45).

Online gaming also involves financial institutions when money is staked. Financial institutions are involved when money is forwarded to an online casino for the purpose of gaming. Attempts have been made to restrict financial transactions associated with child pornography (ACMA, 2009b, p. 69). The Financial Coalition Against Child Pornography has members from banking, payment industries, and internet service companies that seek to target payment systems and obstruct the flow of funds. An industry response has been to seek alternative payment systems. For instance in Australia distributors of prepaid cards have been sought for the pornography industry (www.atlasmedia.net.au).

Where online gaming is illegal, there have been attempts to restrict the movement of funds for the purpose of gambling (Merzer, 2009). America's "wire transfer act" (US Government Accounting Office, 2002, p. 12) has been used for this purpose (Rose, 2006). In the USA credit card companies will not engage in transactions involving gambling (Merzer, 2009). This is partly due to concern as to the legality of transactions but also the risk to the consumer (US Government Accounting Office, 2002). Credit card companies such as American Express will not accept internet gambling providers as appropriate merchants (US Government Accounting Office, 2002, pp. 20-21). Credit card associations such as Visa and Mastercard assign codes to transactions. Transactions coded as internet gambling are to be blocked (Merzer, 2009).

Money laundering is potentially a concern for governments and casinos (Taylor, 2003, pp. 121-130). In addition to restrictions on the flow of money, there may be a variety of obligations imposed by governments to record and monitor the flow of money (ASIC - http://www.asic.gov.au/asic/asic.nsf), and information (ACMA - http://www.acma.gov.au/WEB/HOMEPAGE/PC=HOME). For legal issues associated with the monitoring of internet traffic see Branch (2003).

With respect to issues such as financial transactions, unfortunately there have been problems coding and determining the legality of transactions (US Government Accounting Office, 2002, p. 22). In addition, the audit trail may be confused by the involvement of third parties (Taylor, 2003, pp. 121-130). Debit cards (Owens, 2006) or electronic cash systems such as PayPal are potentially a means of disguising gambling transactions (US Government Accounting Office, 2002, p. 27), but the USA has been pursuing such companies for supporting illegal activities (see http://www.usdoj.gov/usao/nys/pressreleases/January07/Neteller%20Arrests%20PR.pdf Accessed 21/10/2010).

BIOMETRICS

Biometrics has been described as "...the science of establishing the identity of an individual based on the physical, chemical or behavioral attributes of the person" (Jain, Flynn, and Ros[s], 2008). Such attributes include a person's signature (Lee, Berger, and Aviczer, 1996), fingerprints (Maltoni, Maio, Jain, and Prabhakar, 2003), palm-prints (Zhang, 2004), iris pattern (Daugman, 2004), face (Hammoud, Abidi, and Abidi, 2007), gait (Nixon, Tan, and Chellappa, 2006) and possibly even their odour or ear characteristics (European Commission Directorate, 2005).

While many of an individual's attributes remain constant as they age (e.g. fingerprint/palm-print) others change over time (e.g. face, voice and gait) and it is those attributes that have the potential for age authentication. There is however, a significant technical issue associated with using biometrics.

Identification of an individual is easier than verifying that someone is underage. Systems seeking to establish whether someone is underage are not foolproof, and typically incur some degree of error, at times identifying someone as underage when they are not, or identifying someone as overage when then are not. For instance Ni, Song, and Yan (2009) attempted to develop a universal age estimator based upon facial information. They reported absolute errors ranging from 5 to 10 years when attempting to estimate the ages of people in the critical 10-29 year old age. Nevertheless, computer systems appear to be approaching the accuracy of humans in determining the age of an individual from their face (Geng, Zhou, Zhang, Li, and Dai, 2006). Performance may increase when additional biometric measures are used (Jain and Ross, 2004).

Unlike passwords where a perfect match between the user-supplied alphanumeric password and that stored in the system is both required and obtainable for authentication, some allowance for variation must be built into any biometric system to account for factors such as sensor degradation, variations in lighting (important in face recognition for example), changes in biometric characteristics due to illness etc. Allowance for variation results in uncertainty and it is that uncertainty which compromises the accuracy of biometric verification (Jain and Ross, 2009).

It was considerations such as those that led a 2005 European Commission into the impact of biometrics on society to conclude:

"...biometric identification is not perfect - it is never 100% certain, it
is vulnerable to errors and it can be 'spoofed'. Decision-makers need to
understand the level of security guaranteed through the use of biometric
systems and the difference that can exist between the perception and the
reality of the sense of security provided. The biometric system is only one
part of an overall identification or authentication process, and the other parts
of that process will play an equal role in determining its effectiveness."
(European Commission Directorate, 2005).

The "other parts" of an identification/authentication process alluded to by the Commission included information about credit/debit/ID card etc. details. While a combination of biometrics and credit card details might serve to verify the age of an adult user, the provision of credit/debit/ID card details by a minor is a difficulty in the on-line verification of an under-age user.

Any on-line verification process requires information supplied by a user to be compared against information stored in public and private databases. Verification is achieved if an acceptable degree of match is found between stored information and information supplied by the purported user. In the case of minors however, there is little valid information available in databases since society restricts the requirement to have and divulge identifying information to persons over the legal "adult" age. Minors do not, for example, have a drivers license, tax file number, credit card, health care card or own property.

Such considerations were discussed in a 2008 European Commission into, amongst other things, age verification across various media including the internet and mobile platforms. The commission reviewed a number of options and sought input from age verification solutions providers, including those advocating biometrics. The commission found biometrics to be incapable of predicting exact age and ultimately concluded that while some multi-faceted approaches involving for example, electronic identity cards (eID cards) or credit cards showed promise (European Commission Directorate-General, Information Society and Media, 2008):

"...most stakeholders seem to agree that there is no existing approach to Age Verification that is as effective as one could ideally hope for, a view shared by those Age Verification Solution providers and services present at the Safer Internet Forum".

A similar finding was recently reported following a legal battle between the American Civil Liberties Union and the US Attorney General over the Child Online Protection Act (COPA), introduced by the US Congress in 1998 and aimed at controlling children's access to sexually explicit material over the internet. The law was immediately challenged by American Civil Liberties Union on the basis of, amongst other issues, free speech and so instigation of the law was delayed until the matter was resolved in court. Resolution took some 10 years with the matter eventually reaching the US Supreme Court which declared the law unconstitutional on January 22, 2009 (Neuburger, 2009).

Amongst reported findings were :

"...that there is no evidence of age verification services or products available on the market to owners of Web sites that actually reliably establish or verify the age of Internet users. Nor is there evidence of such services or products that can effectively prevent access to Web pages by a minor" (US Attorney General, 2007).

The trial also noted that requiring debit, credit or payment card information was no guarantee of age since the person entering the information does not have to be the person to whom the information applies. Further problems associated with a law requiring age verification technology included: the possible imposition of a cost to content providers owing to lost traffic; loss of anonymity to users; and jurisdictional considerations that prevent enforcement of any such requirement on Websites located outside the USA (US Court of Appeals, 2008) thereby allowing determined minors continued access to inappropriate material.

FILTERS

Content filters are another attempt to control the information that is available on the internet. Filters can either operate on a black-list principle or a white-list principle. A Black-list filter allows access to all but a specific list of sites (http://www.acma.gov.au/WEB/STANDARD/pc=PC_90167). It is more flexible, but vulnerable to problems as the list of inappropriate sites requires updating, for example when the blacklist is leaked (see Moses, 2009). A White-list filter blocks access to all but a specific list of sites. It is far more secure, but severely curtails the material that can be accessed using the internet (see Peltz, 2002). The Australian government has attempted to implement content filters (at a cost of $84M) to assist the censoring of internet content (http://www.netalert.gov.au/filters.html), but with limited success (ACMA, 2009b). There is a limited uptake of content filters, possibly because the content filters were difficult to install (ACMA, 2009b), and the content filters could be circumvented (Turner, 2008).

In the UK, Byron (2008) suggested that content filters be loaded on all commercially available computers at point of sale. Alternatively government sponsored content filters could be installed automatically with internet accounts, as has been arranged in France (ACMA,

2009b). For instance the internet subscribers in France who choose to have parental controls installed select one of three standard profiles: 1) the *child profile* allows access only to a pre-defined index of websites; 2) the *teenager profile* uses indexes of pre-assessed content to block access to online gambling and internet content that is sexually explicit, promotes hatred or violence or is drug related; 3) the *adult profile* allows open access to internet content. Other systems developed in the US seek to impose time limits on access (ACMA, 2009b, p. 38). In the UK, providers of mobile phones have voluntarily acted to restrict access to internet content (http://www.imcb.org.uk/assets/documents/10000109Codeofpractice.pdf). The Code covers new types of content, including visual content, online gambling, mobile gaming, chat rooms and Internet access. It does not cover traditional premium rate voice or premium rate SMS (texting) services, that continue to be regulated under previous codes of practice. The European Union also has codes that make provisions for restrictions upon mobile phone content and times of internet access.

Nevertheless, there are problems with the use of white or blacklist filters to block access to inappropriate sites. There can be discussion as to which sites are inappropriate. For instance content filters have been used to censor access to "inappropriate" sites, such as those criticising the Thai royal family (Moses, 2009). Although considerable effort has been devoted to developing and updating filters they remain prone to error: missing inappropriate sites; and blocking appropriate sites (Gedda, 2009a,b; Moses, 2008; Peltz, 2002). For example, there was complaint when the UK company Betfair was one of the sites included in the Australian government's "black list" (Pauli, 2009).

Unlike the blacklist approach, other classes of filters operate on categories of content or keywords (http://www.netalert.gov.au/filters/faqs.html#q6), but are still prone to some degree of error. For instance, Ho and Watters (2004) claimed better performance (99.1%) filtering pornography sites, by filters that analysed the structure of the websites and detecting key words used. Chou, Sinha, and Zhao (2008) report similar success rates detecting inappropriate internet use within a specific workplace (computer programming in the IT industry) using text mining and text categorisation. Even so, because these filters target a specific application or domain, they are less likely to be successful across multiple applications and domains.

Hunter (2000) compared four commercially available filters and examined their ability to filter 200 websites. Of these websites 18% had objectionable content (language, nudity, sex or violence). On average these filters correctly blocked objectionable material 75% of the time, but also excluded 21% of non-objectionable material. Peltz (2002) reports an 80% accuracy is common for filtering technology. Nevertheless, some of the anecdotally reported errors involve the blocking of the sites belonging to filtering campaigners, an irony that subjected proponents to potential ridicule (Peltz, 2002, p. 413). The Australian Library and Information Association (2007) reported the results of a survey on internet filtering. Of the 104 responding libraries 39% used filtering software, but were concerned as to the reliability and accuracy of filtering software. Although these libraries primarily filtered pornography, violence, hate, and web-based mail, libraries also filtered gambling.

There would be other options where internet gaming is legalised that utilise the architecture of the internet. Currently voluntary standards have been created for regulating internet casinos (Scoolidge, 2006). An organisation called the e-Commerce and Online Gaming Regulation and Assurance (eCOGRA) have suggested a set of standards called eCOGRA Generally Accepted Practices (eGAP). The standards set forth by eCOGRA address money laundering, fraud, underage and problem gambling (Scoolidge, 2006). Organisations that comply with these voluntary standards can claim a seal of approval. Failure to comply with such standards can result in the removal of approval. Admittedly this is unlikely to do much to deter a disreputable operator. Hence Scoolidge (2006) suggested that approved operators be registered on a specific internet domain name. On the internet, domain names serve as registries and methods of navigating on the internet. Domain names such as .COM or .ORG indicate a commercial company or organisation. Scoolidge (2006) suggested that a specific domain be created for approved and regulated internet gaming, such that operators that did not comply with approved standards could be removed from this approved domain. Gamblers cannot place money on a website they cannot see.

SURVEILLANCE

Adult supervision is likely to reduce access to inappropriate content (Rockloff and Greer, in press; Rosen, Cheever, and Carrier, 2008; Wang, Bianchi, and Raley, 2005), but technological solutions are also available. The history of browsers can be accessed, and there are methods of monitoring the activity of mobile phones (http://www.flexispy.com/). Dedicated child surveillance software is available (http://www.kidtracker.com.au/)

If consumers gamble electronically, there is the potential for them to be tracked (Shaffer, Peller, LaPlante, Nelson, and LaBrie, 2010) and either offered inducements (Austin, and Reed, 1999; Browne, 2005; Sévigny, Cloutier, Pelletier, and Ladouceur, 2005) and advice (Dedonno and Detterman, 2008), or alternatively given electronic warnings (Monaghan, 2009; Monaghan, Derevensky, and Sklar, 2008) to minimise harm. Although there is the potential with mobile phone technology for individuals to gamble anywhere at any time, there is also the potential to set limits (Broda et al., 2008; Nelson et al, 2008; Nower and Blaszczynski, 2010), provide consumer warnings (Cloutier, Ladouceur, and Sévigny, 2006; Floyd, Whelan, and Meyers, 2006; Monaghan and Blaszczynski, 2007) and to target specific individuals (using biometrics) at specific locations (Aalto, Göthlin, Korhonen, and Ojala, 2004; Yang, Cheng, and Dia, 2008) and times.

It is also technically possible to indiscriminately block access to inappropriate content. As mobile phones are radios, transmissions can be intercepted by scanners (http://www.spyequipmentguide.com/cell-phone-scanners.html) or jammed by broadcasting on similar frequencies (http://en.wikipedia.org/wiki/Mobile_phone_jammer). However mobile phone jammers tend to be illegal in many jurisdictions, and are more appropriately used by security forces or in prisons. France attempted to jam mobile phones at their Opera house, but this was discontinued because the jamming was indiscriminate and interfered with the use of mobile phones for emergency purposes (http://www.fiercemobileit.com/story/problem-jamming- mobile-phone/2009-05-13).

Specific websites can be targeted instead. Consumer access to internet gambling sites has been blocked by criminal groups making extortion demands (Heath, 2008). The websites of organisations not complying with demands for money are swamped by multiple (ghost) attempts to access their website (Leyden, 2004). The fictitiously high traffic on the server denies service to genuine consumers. These are called Denial of Service attacks. Threats of this nature were made during the Melbourne cup in 2007 (Tung, 2007). Owens (2006) suggests that this is one of the areas where regulators could assist the providers of online gambling.

CONCLUSION

The long term goal of organisations regulating the internet is to allow access to electronic services to all individuals using any web capable platform. Hence this chapter considered likely mechanisms that could control access to inappropriate content such as pornography or gambling. Filters can be used to block access to pornography or gambling, but they are unlikely to be 100% successful. In addition, biometrics alone will not suffice to block underage access to inappropriate content. Nevertheless a combination of these approaches may be more efficacious.

The difficulty in enforcing national laws internationally means there are few legal consequences to operators who provide minors with access to inappropriate content. That fact, in conjunction with technical limitations and a potential loss in revenue should minors be successfully prohibited from accessing their sites, currently provides site owners both little impetus or ability to prevent minors accessing age-inappropriate content. Although attempts are being made to restrict access to inappropriate content, it is a complex problem that is unlikely to be solved by legislators in the near future.

ACKNOWLEDGMENTS

The authors would like to acknowledge funding support from Gambling Research Australia (Tender No 119/06).

REFERENCES

Aalto, L. Göthlin, N., Korhonen, J., and Ojala, T. (2004). *Bluetooth and WAP push based location-aware mobile advertising system.* In MobiSys '04: Proceedings of the 2nd international conference on Mobile systems, applications, and services (2004), pp. 49-58.

Australian Bureau of Statistics (2008-09). *Household Use of Information Technology, Australia, 2008-09,* Cat No, 8146.0 (Available at http://abs.gov.au/AUSSTATS/abs@.nsf/Lookup/8146.0Main+Features12008-09?OpenDocument)

ACMA, (2008). *Australian Communications and Media Authority.* Communications Report 2007-2008.

ACMA, (2009a). Mobile phone technology and special features fact sheet.

ACMA (2009b). *Developments in internet filtering technologies and other measures for promoting online safety.* [Downloaded 27th April 2009 from http://www.acma.gov.au/webwr/_assets/main/lib310554/developments_in_internet_filters_2ndreport.pdf]

Australian Library and Information Association (2007*). Internet filtering in public libaries 2007 survey report.* [http://alia.org.au/advocacy/internet.access]

Austin, M.J., and Reed, M.L. (1999). Targeting children online: Internet advertising ethics issues. *Journal of Consumer Marketing, 16*(6), 590-602.

Boyera, S. (2006). The mobile web to bridge the digital divide. [downloaded 13th May from *http://www-mit.w3.org/2006/12/digital_divide/IST-africa-final.pdf]*

Branch, P. (2003). Lawful interception of the internet. *Australian Journal of Emerging Technologies and Society, 1*(1), 38-51.

Broda, A., LaPante, D.A., Nelson, S.E., LaBrie, R.A., Bosworth, L.B., and Shaffer, H.J. (2008). Virtual harm reduction efforts for Internet gambling: effects of deposit limits on actual Internet sports gambling behavior. *Harm Reduction Journal, 5,* 27-36.

Browne, S. (2005). *The math of player development.* Raving Consulting Press: Reno, Nevada.

Byron, T. (2008) *Safer children in a digital world.* [Downloaded 27th April 2009 from *http://www.dcsf.gov.uk/byronreview/pdfs/Final%20Report%20Bookmarked.pdf*

Chou, C-H., Sinha, A.P., and Zhao, H. (2008). A text mining approach to Internet abuse detection. *Information Systems and E-Business Management, 6,* 419-439.

Cloutier, M., Ladouceur, R., and Sévigny, S. (2006). Responsible gambling tools" Pop-up messages and pauses on video lottery terminals, *The Journal of Psychology, 140*(5), 434-438.

Daugman, J. (2004). How Iris Recognition Works? *IEEE Transactions on Circuits and Systems for Video Technology, 14*(1), 21–30.

Dedonno, M.A., and Detterman, D.K. (2008). Poker is a skill. *Gaming Law Review, 12*(1), 31-36.

Eadington, W.R. (1988). Loophole legalization: The process of the spread of commercial gambling without popular or legislative intent. (pp. 1- 12). In M. Dickerson (Ed.), *200-up* (pp. 1-12*). Proceedings of the 3rd National Conference of the National Association for Gambling Studies*, Canberra.

Eadington, W.R. (2004). The future of online gambling in the United States and elsewhere. *Journal of Public Policy and Marketing, 23*(2), 214-219.

European Commission Directorate-General. Joint Research Centre. Institute for Prospective Technological Studies (2005). *Biometrics at the Frontiers: Assessing the Impact on Society For the European Parliament Committee on Citizens' Freedoms and Rights, Justice and Home Affairs* (LIBE). P. 10. Available at :
http://ec.europa.eu/justice_home/doc_centre/freetravel/doc/biometrics_eur21585_en.pdf

European Commission Directorate-General. Information Society and Media (2008): *Background Report on Cross Media Rating and Classification and Age Verification Solutions. Safer Internet Forum.* Sept 25-26, Luxembourg. P. 26. Available at : *http://ec.europa.eu/information_society/activities/sip/docs/pub_consult_age_rating_sns/r eportageverification.pdf*

Finn, M. (2005). Gaming goes mobile: Issues and implications. *Australian Journal of Emerging Technologies and Society, 3*(1), 31-42.

Floyd, K., Whelan, J.P., and Meyers, A.W. (2006). Use of warning message to modify gambling beliefs and behavior in a laboratory investigation. *Psychology of Addictive Behaviors, 20*(1), 69-74.

Gedda, R. (2009a). *Banned poker sites make joker of ACMA's Internet blacklist.* [downloaded 28th April 2009 from
http://www.arnnet.com.au/article/296460/banned_poker_sites_make_joker_acma_intern et_blacklist]

Gedda, R. (2009b). *URL blacklist 'creep' possible: Conroy.* [downloaded 28th April 2009 from http://www.computerworld.com.au/article/296842/url_blacklist_creep_possible_conroy]

Geng, X., Zhou, Z-H., Zhang, Y., Li, G., and Dai, H. (2006). Learning from facial aging patterns for automatic age estimation. *Proceedings of the 14th annual ACM International Conference on Multimedia*, pp. 307-316.

Griffiths, M.D. (2003). Internet gambling: Issues, concerns and recommendations. *CyberPsychology and Behavior, 6*, 557-568.

Griffiths, M. (2007). Interactive television quizzes as gambling: A cause for concern? *Journal of Gambling Issues, 20*, 269-76.

Hammoud, R.I., Abidi, B.R., and Abidi, M.A. (Eds.) (2007). *Face biometrics for personal identification multi-sensory multi-modal systems.* New York : Springer.

Heath, N. (2008). *Gambling site brought to its knees by 'unstoppable' botnet.* [downloaded 30th April from http://software.silicon.com/security/0,39024655,39170296,00.htm]

Ho, W.H., and Watters, P.A. (2004). Statistical and structural approaches to filtering internet pornograpy. *IEEE International Conference on Systems, Man and Cybernetics*, 4792-4798.

Hunter, C.D. (2000). Internet filter effectiveness - testing over- and underinclusive blocking decisions of four popular web filters. *Social Science Computer Review, 18*, 214-222.

ITU, 2003. The birth of broadband. Media Fact Sheet. International Telecommunication Union. September. Retrieved 12 June, 2009 from *http://www.itu.int/osg/spu/publications/birthofbroadband/faq.html*

Jain, A.K,, Flynn, P., and Ross, A.A. (Eds). (2008). *Handbook of biometrics.* Boston, MA : Springer Science and Business Media.

Jain, A.K., and Ross, A.A. (2004). Multibiometric systems. *Communications of the ACM, 47*(1), 34-40.

Jain, A.K. and Ross, A.A. (2009). *Introduction to biometrics.* Berlin: Springer

Kraut, R.E., Sunder, S., Telang, R., and Morris, J. (2005). Pricing electronic mail to solve the problem of spam. *Human Computer Interaction, 20*, 195-223.

Lee, L., Berger, T and Aviczer, E (1996). Reliable on-line human signature verification systems. *IEEE Transactions on Pattern Analysis and Machine Intelligence, 18*(6), pp. 643–647

Leyden, J. (2004). Extortionists take out UK gambling site. [downloaded 30th April from *http://www.theregister.co.uk/2004/04/05/sporting_options_ddosed/*]

Mahatanankoon, P., Wen, H.J., and Lim, B. (2005). Consumer-based m-commerce: exploring consumer perception of mobile applications. *Computer Standards and Interfaces, 27*,

Maltoni, D., Maio, D., Jain, A.K., and Prabhakar, S (2003). *Handbook of fingerprint recognition.* Berlin: Springer-Verlag,

Media release, 2009. New National Broadband Network. Office of the Prime Minister of Australia. 7 April 2009. Retrieved 11 June, 2009 from *http://www.pm.gov.au/media/Release/2009/media_release_0903.cfm*

Merzer, M. (2009). *New Internet gambling regulations go into effect.* [downloaded 29th April from http://www.creditcards.com/credit-card-news/unlawful-internet-gambling-enforcement-act-credit-card-1282.php]

Monaghan, S. (2009). Responsible gambling strategies for internet gambling" The theoretical and empirical base of using pop-up messages to encourage self-awareness. *Computers in Human Behavior, 25*, 202-207.

Monaghan, S. and Blaszczynski, A. (2007). Recall of electronic gaming machine signs: A static versus a dynamic mode of presentation. *Journal of Gambling Issues, 20*, 253- 267.

Monaghan, S., Derevensky, J., and Sklar, A. (2008). Impact of gambling advertisements and marketing on children and adolescents: Policy recommendations to minimise harm. *Journal of Gambling Issues, 22*, 252-274.

Moses, A. (2009). *Leaked Australian blacklist reveals banned sites.* [downloaded 29th April, 2009 from http://www.smh.com.au/articles/2009/03/19/1237054961100.html

Moses, A. (2008). *Net filters may block porn and gambling sites.* [downloaded 29th April , 2009 from http://www.smh.com.au/articles/2008/10/27/1224955916155.html]

Nelson, S.E., LaPlante, D.A., Peller, A.J., Schumann, A., LaBrie, R.A., and Shaffer, H.J. (2008). Real limits in the virtual world: Self-limiting behavior of internet gamblers. *Journal of Gambling Studies, 24*, 463-477.

Neuburger, J.D (2009): *U.S. Supreme Court (Finally) Kills Online Age Verification Law.* Mediashift. January 29. Available at: *http://www.pbs.org/mediashift/2009/01/us-supreme-court-finally-kills-online-age-verification-law029.html.* Accessed: 30/4/09

Ni, B., Song, Z., and Yan, S. (2009). Web image mining towards universal age estimator. *Proceedings of MM'09, Oct 19-24, 2009*, Beijing, China.

Nixon, M.S., Tan, T.N., and Chellappa, R. (2006). *Human identification based on gait.* London: Springer.

Nower, L., and Blaszczynski, A. (2010). Gambling motivations, money-limiting strategies, and precommitment preferences of problem versus non-problem gamblers. *Journal of Gambling Studies.*

Owens, M.D.jr. (2006). If you can't beat 'em, will they let you join? What American states can offer to attract internet gambling operators. *Gaming Law Review, 10*(1), 26-32.

Parke, A., and Griffiths, M.D. (2004). Why internet gambling prohibition will ultimately fail. *Gambling Law Review, 8*(5), 295-299.

Pauli, D. (2009). *Betfair banned by ACMA.* [downloaded from *http://www.arnnet.com.au/article/296165/betfair_banned_by_acma]*

Peltz, R.J. (2002). Use "the filter you were born with": The unconstitutionality of mandatory internet filtering for the adult patrons of public libraries. *Washington Law Review, 77*, 397-479.

Productivity Commission (2010). Gambling: Public inquiry. Accessed via *http://www.pc.gov.au/projects/inquiry/gambling-2009 on 7/9/10.*

Rockloff, M.J., and Greer, N. (in press). Audience influence on EGM gambling: the protective effects of having others watch you play. *Journal of Gambling Studies.* Doi:10.1007/s10899-010-9213-1.

Rose, I.N. (2006). *The Unlawful Internet Gambling Enforcement Act of 2006 analyzed.* [downloaded 29th April 2009 from *http://www.casinocitytimes.com/article.cfm?ContentAndContributorID=30106]*

Rosen, L.D., Cheever, N.A., and Carrier, L.M. (2008). The association of parenting style and child age with parental limit setting and adolescent MySpace behavior. *Journal of Applied Developmental Psychology, 29*, 459-471.

Scoolidge, P.J. (2006). Gambling blindfolded: The case for a regulated domain for gambling sites. *Gaming Law Review, 10*(3), 252-265.

Sévigny, S., Cloutier, M., Pelletier, M-F., and Ladouceur, R. (2005). Internet gambling: Misleading payout rates during the "demo" period. *Computers in Human Behavior, 21*, 153-158.

Shaffer, H.J., Peller, A.J., LaPlante, D.A., Nelson, S.E., and LaBrie, R.A. (2010). Toward a paradigm shift in Internet gambling research: From opinion and self-report to actual behavior. *Addiction Research and Theory, 18*(3), 270-283.

Taylor, R.J. (2003). *Casino crimes and scams* (pp. 121-130). New York: Vantage Press.

Tung, L. (2007). *NAB floats denial-of-service threats to the cloud.* [downloaded 30th April 2009 from http://www.zdnet.com.au/news/security/soa/NAB-floats-denial-of-service-threats-to-the-cloud/0,130061744,339284465,00.htm]

Turner, A. (2008). *Australian internet filters have backdoor. Australian PC Authority.* [downloaded 28th April 2009 from *http://www.pcauthority.com.au/BlogEntry/127342,australian-internet-filters-have-backdoor.aspx*]

US Attorney General (2007). Civil Action No. 98-5591 in the United States District Court of Pennsylvania. Final Adjudication. *American Civil Liberties Union* et al. *v Alberto R. Gonzales,* US Attorney General. March 22, 2007. paragraph 138.

US Court of appeals No. 07-2537. *American Civil Liberties Union* et al. *v Michael B. Mukasey, US Attorney General. Argued*: June 10 2008, Filed: July 22 2008.

US-FCC, 2009. What is broadband ? Site reviewed/updated 27/5/09. Retrieved 12 June, 2009 from *http://www.fcc.gov/cgb/broadband.html*

US Government Accounting Office (2002). Internet gambling: An overview of the issues. [downloaded 24th April 2009 from *http://www.gao.gov/new.items/d0389.pdf*]

Wang, R., Bianchi, S.M., and Raley, S.B. (2005). Teenagers' internet use and family rules: A research note. *Journal of Marriage and Family, 67*, 1249-1258.

Watson, S., Liddell, P.Jr., Moore, R.S., and Eshee, W.D. Jr. (2004). The legalisation of internet gambling: A consumer protection perspective. *Journal of Public Policy and Marketing, 23*(2), 209-213.

Yang, W.-S., Cheng, H.-C., and Dia, J.-B. (2008). A location-aware recommender system for mobile shopping environments. *Expert Systems with Applications, 34*(1), 437-445.

Zhang, D.D. (2004). *Palmprint authentication.* Norwell, Mass.: Kluwer Academic Publishers.

INDEX

2

20th century, 57, 159
21st century, 156, 162, 172, 177

3

3D acceleration, x, 192

A

access, x, 57, 58, 60, 74, 84, 87, 88, 90, 147, 148, 149, 184, 188, 227, 228, 229, 230, 231, 232, 235, 236, 237, 238
accessibility, 66, 75
acetylcholinesterase, 100
acid, 124
acoustic neuroma, 101, 117, 127
acquisitions, 105, 172, 175, 181
adaptability, 120
adaptations, 75, 76
adenine, 102
adolescents, x, 227, 241
adults, 100, 125, 129, 184, 186, 232
advancement, 96, 120
advertisements, 241
advocacy, 238
Africa, 177, 180
age, viii, x, 69, 75, 77, 81, 86, 88, 95, 102, 104, 106, 107, 118, 124, 184, 186, 227, 230, 231, 232, 233, 234, 235, 238, 239, 240, 241
ageing population, 184
agencies, 103
agility, 137, 138, 148
airports, 76
alcohol consumption, 118, 129, 185
alcoholics, 107

American Civil Liberties Union, 235, 242
amine, 122
amplitude, 37, 38, 40, 42
anemia, 127
animations, 56, 84, 139
ANOVA, 6, 7, 106
apoptosis, 21, 27, 28, 30, 43, 52, 53, 97, 99, 121, 125, 127, 131
application component, 140
aqueous solutions, 41, 46
Argentina, 58, 177, 178, 179
arrest, 104
Asia, 78, 177, 180
assessment, 51, 65, 66, 97, 124, 128
assets, 236, 238
audit, 231, 233
Austria, 156, 160, 161, 169, 170, 173
authentication, 231, 233, 234, 242
autism, 57
automate, 64
automation, 84
autonomy, 144, 151
Azerbaijan, 178, 179

B

background radiation, viii, 95, 96, 100, 101, 127
ban, 228
bandwidth, 228
Bangladesh, 178, 179
banking, 76, 184, 232
barriers, 81
base, viii, 2, 3, 8, 11, 12, 17, 18, 20, 27, 33, 34, 35, 47, 48, 50, 51, 102, 103, 107, 113, 117, 123, 125, 130, 164, 180, 187, 188, 236, 241
batteries, 56, 76, 77, 79, 82
Beijing, 152, 153, 241
Belgium, 156, 161, 162, 169, 171, 174, 175, 191

benefits, 130, 148, 150, 151, 187, 227
beta-carotene, 129
binucleated cells (BN), viii, 95, 104
biochemical action, 3, 45
biochemical processes, 37
biological activity, 9
biological models, vii, 1, 3
biological processes, 3
biological responses, 103
biological systems, 98
biomarkers, viii, 95, 97, 104, 119
Biometric systems, xi, 227
biomolecules, 44, 99
biomonitoring, 97, 102, 105, 124, 125
birds, 36
birth rate, 184
blood, 18, 100, 102, 104, 114, 117, 127, 129, 130, 131, 184, 185, 186
blood pressure, 184, 185, 186
blood-brain barrier, 18
Bluetooth, x, 60, 61, 76, 185, 238
body mass index, 189
Boltzmann constant, 41
bonds, vii, 1, 43, 44
bone, 117, 125
bone marrow, 117, 125
Botswana, 177, 178
brain, 3, 35, 36, 44, 49, 50, 51, 52, 53, 100, 101, 117, 118, 119, 124, 126, 127, 130
brain cancer, 44, 100, 117, 119, 126
brain tumor, 36, 51, 117, 124
Brazil, 133
breast cancer, 130
breeding, 50
buccal mucosa, 104, 108, 109, 110, 118, 128, 129
businesses, 156
buttons, 83
bystander effect, 100, 121, 127

C

calcium, 43, 49, 102
calibration, x, 105
caloric intake, 186
Cameroon, 177, 178, 179
cancer, viii, 2, 3, 36, 44, 50, 95, 98, 99, 100, 102, 104, 105, 107, 117, 118, 119, 122, 124, 125, 126, 127, 129, 130
carcinogen, viii, 95, 98
carcinogenesis, 97, 102, 122, 125
caregivers, 187
Caribbean, 177, 178
carotene, 129

case studies, 151
casinos, 233, 236
categorization, 98
cation, 38, 39, 40, 41, 42, 43
cell culture, 119
cell cycle, 99, 118
cell death, 2, 21, 22, 23, 24, 25, 26, 27, 28, 29, 30, 34, 35, 36, 43, 45, 49, 50, 52, 99, 101
cell differentiation, 110
cell line, 50, 126
cell lines, 126
cell membranes, vii, 2, 3, 18, 35, 36, 39, 41, 43, 44
cell organelles, 98, 129
cell phones, 58, 103, 185
Central African Republic, 177, 178
central nervous system, 49, 100
chain molecules, 122
challenges, 150, 188, 228
chemical, vii, 1, 21, 27, 35, 43, 96, 102, 104, 128, 233
chemical bonds, vii, 1, 43
chemicals, 27, 100, 126, 128
chemopreventive agents, 128
chicken, 35, 49
childhood, 99, 129, 130
children, x, 48, 86, 88, 99, 100, 117, 184, 186, 189, 227, 235, 239, 241
Chile, 58, 177, 178
China, 82, 152, 153, 241
chromatid, 97, 118
chromosome, 97, 102, 105, 117, 118, 121, 124
circulation, 67
CIS, 127
citizens, 150
classes, 114, 236
classification, 67
classroom, 135, 137, 151
CNS, 124
coding, 233
cognitive performance, 51
collaboration, ix, 133, 135, 137, 138, 139, 140, 141, 143, 144, 145, 146, 147, 148, 149, 150, 151
college students, 188
Colombia, 177, 178, 179
colonization, 179
color, 59, 60
combined effect, 119
Comet Assay, viii, 95, 105, 123, 129
commerce, 62, 65, 85, 240
commercial, 56, 58, 59, 75, 76, 80, 81, 88, 158, 171, 237, 239
communication, vii, viii, ix, 11, 56, 57, 58, 61, 62, 66, 75, 81, 83, 85, 88, 95, 96, 99, 101, 102, 124,

133, 134, 135, 137, 138, 139, 143, 144, 148, 150, 184, 185, 186, 229
communication systems, 57, 96, 99
communication technologies, 61, 185
communities, ix, 60, 64, 70, 75, 133, 135, 136, 138, 139, 141, 148, 150, 151, 186, 187, 188, 189
community, vii, x, 56, 83, 88, 134, 136, 138, 148, 150, 186, 187, 188, 227, 230
community support, 188
compass measurement, x, 192, 204
compatibility, 62, 74, 75, 83, 156, 180
competition, 62, 181, 230
competitors, 83, 172
compilation, 67, 69, 165, 170
complexity, 57
complications, 80, 121
compounds, 104
computer, ix, 56, 57, 61, 69, 74, 78, 81, 83, 84, 86, 89, 106, 122, 130, 133, 137, 148, 149, 152, 153, 187, 229, 230, 234, 236
computer software, 106
computer systems, 187, 234
computer technology, 229
computing, x, 47, 56, 59, 67, 69, 70, 74, 76, 82, 84, 90, 134, 135, 148, 150, 152, 229
conception, 75, 79, 81, 84, 135
conductivity, 37, 46, 47
conference, 152, 153, 238
connectivity, 74, 149, 228
consensus, 77, 148
consent, 104
consolidation, 143
constituents, 17
construction, 136, 137
consumer protection, 242
consumers, 172, 229, 237
consumption, 78, 79, 82, 84, 118, 129, 185
contradiction, 119
control group, ix, 96, 119, 120
controversial, 119
convergence, 64, 80
conversations, 48, 60, 62, 143
coordination, 67, 128, 137
copper, 228
copyright, iv
Copyright, iv
correlation, ix, 16, 18, 96, 103, 106, 111, 113, 116, 118, 119, 120, 124, 139
correlation analysis, 16, 18, 119
correlation coefficient, 16
cost, x, 56, 60, 62, 64, 66, 67, 70, 73, 77, 81, 82, 83, 150, 158, 159, 230, 232, 235
cotton, 4

country of origin, 232
Court of Appeals, 235
CPU, 59, 60
creativity, 62, 83, 148
creep, 240
crimes, 242
critical value, 37, 38
CRM, 84
Croatia, 178, 179
cultural practices, 138
culture, 4, 82, 100, 119
currency, 63
current limit, 47
customers, 81, 173
cytokinesis, 124
cytoplasm, 37, 39, 43, 46
cytosine, 102
cytoskeleton, 28, 29, 30, 34
cytostatic drugs, 104
Czech Republic, 156, 160, 161, 163, 165, 169, 170, 171, 172, 173, 174, 175

D

damping, 37
data analysis, 105
data collection, 80, 93
deaths, 49
defence, 27
deficiency, 50, 128
degenerate, 27
degradation, 234
democracy, 152
democratisation, 227
democratization, 57, 63
denial, 242
Denmark, 100, 125, 156, 161, 162, 163, 164, 169, 171, 172, 173, 174, 175, 189
depression, 186, 189
designers, 56, 79, 88
detection, x, 22, 29, 97, 121, 123, 124, 128, 129, 130, 239
developed countries, 59, 64, 228
developed nations, 227
developing countries, 64, 96, 188
developmental process, 4
deviation, 9, 10, 12, 15, 17, 26, 33, 73, 74, 114
diabetes, 184, 186, 188
diet, 184, 186
dietary habits, 106, 119
dielectric constant, 39, 42, 46
diet, 184, 186
dietary habits, 106, 119
diffusion, 58, 64, 157, 188

digital divide, x, 227, 239
digital television, 76
diploid, 100, 119
direct action, 43
direct observation, 88
directors, 63
discs, 70
diseases, viii, 51, 95
dispersion, 74
displacement, 37, 38, 39, 40, 41, 42
distance education, 134
distortions, 73
distribution, 46, 73, 82, 158, 169
diversification, ix, 155, 156, 163
DNA, vii, viii, 1, 2, 3, 21, 22, 23, 24, 25, 26, 27, 28, 29, 30, 34, 35, 36, 43, 44, 45, 49, 50, 51, 52, 53, 54, 95, 97, 98, 99, 100, 101, 102, 104, 105, 106, 108, 114, 115, 116, 117, 118, 119, 120, 121, 122, 123, 124, 125, 126, 127, 128, 129, 130
DNA breakage, 50, 123
DNA damage, vii, viii, 1, 2, 3, 21, 22, 24, 25, 26, 27, 28, 34, 35, 36, 44, 45, 49, 50, 52, 53, 54, 96, 97, 98, 99, 101, 102, 103, 106, 114, 115, 116, 117, 119, 120, 121, 122, 123, 124, 125, 126, 127, 128, 129, 130
DNA lesions, 121
DNA repair, 97, 102, 104, 120, 123, 126, 127
DNA strand breaks, 52, 99, 102, 114, 117, 126
DOI, 126, 128
domestic markets, 159, 180
dominance, 57
Dominican Republic, 177, 178
downlink, 11
Drosophila, 2, 3, 4, 6, 8, 10, 12, 19, 21, 24, 27, 32, 50, 51, 52, 53, 101
drugs, 104, 125
dynamism, 181

E

e-commerce, 62, 65, 85
economic development, 57
economic status, viii, 95, 104
Ecuador, 177, 178
education, ix, 62, 65, 67, 73, 133, 134, 135, 136, 137, 138, 139, 140, 141, 148, 150, 151, 188, 189
educational objective, 136
educational process, 138
educational research, 72
egg, 4, 21, 22, 23, 24, 25, 26, 27, 28, 29, 30, 51, 93, 101, 102, 104, 108, 110, 112, 119, 120
Egypt, 53, 177, 178, 179

El Salvador, 177, 178
elaboration, 67, 69, 160, 162, 163, 164, 165, 166, 167, 168, 172, 175, 176, 177, 178, 180
e-learning, 85, 151
electric charge, 38
electric current, 46
electric field, 5, 9, 15, 16, 17, 18, 36, 37, 38, 40, 42, 46, 103, 122
electricity, 96
electrodes, 61
electromagnetic, vii, viii, 2, 26, 27, 28, 37, 41, 44, 45, 47, 48, 49, 50, 51, 52, 54, 58, 95, 96, 97, 98, 99, 100, 101, 102, 111, 120, 121, 122, 126, 127, 128, 129, 130
electromagnetic fields, vii, 2, 47, 49, 50, 51, 52, 54, 96, 99, 120, 127, 128, 129, 130
electromagnetic fields (EMFs), vii, 2, 99
electromagnetic waves, 49, 97
electromagnetism, 48
electron, 44, 125
electronic circuits, 48
electrons, 43, 44
electrophoresis, 97, 104, 118, 121, 123, 127, 129
e-mail, 81
embryogenesis, 3
emergency, 48, 149, 184, 185, 237
EMG, 89
emission, 5, 6, 9, 12, 30
energy, 30, 45, 46, 47, 79, 82, 97, 99, 100, 103, 117, 122
energy consumption, 79, 82
energy density, 45
enforcement, 235, 240
engineering, 65, 67, 74, 83, 85
England, 52, 54, 181
enlargement, 75
environment, viii, x, 41, 47, 48, 62, 68, 75, 80, 81, 90, 95, 96, 98, 135, 137, 139, 140, 141, 143, 144, 145, 146, 147, 148, 149, 150, 152
environmental conditions, 12
environmental effects, 124
environmental factors, 96
Environmental Protection Agency, 100
environmental stress, 3
enzyme, 100
enzymes, vii, 1, 3, 43
EPA, 100
epidemic, 184
epidemiology, 186
epigenetic alterations, 100
epithelial cells, 52, 54, 101, 104, 105, 108, 109, 110, 123, 126, 128, 129
Equatorial Guinea, 177, 178

equipment, 85
ergonomics, 74
Estonia, 178, 179
ethics, 239
Europe, ix, 3, 56, 64, 76, 78, 86, 89, 128, 155, 156, 157, 158, 159, 163, 164, 165, 166, 169, 172, 175, 176, 179, 180, 181, 186, 188
European Commission, 233, 234, 239
European Community, 128
European market, 159, 163, 172, 176
European OECD countries, ix, 155, 156, 159, 160, 176, 177, 179, 180, 181
European Parliament, 239
European Union (EU), 63, 81, 158, 181, 236
everyday life, vii, viii, ix, 95, 99, 133, 136, 149
evidence, 44, 47, 51, 118, 120, 124, 127, 169, 235
evolution, vii, ix, 48, 56, 57, 58, 59, 66, 74, 76, 77, 79, 80, 82, 89, 122, 134, 150, 155, 156, 157, 159, 163, 169, 181
excision, 102, 124, 125
execution, 67, 70
exercise, 186
exploitation, 139, 158
exposure, vii, ix, 2, 3, 4, 5, 6, 8, 9, 11, 12, 13, 18, 22, 27, 28, 30, 31, 32, 33, 34, 35, 36, 44, 45, 47, 48, 50, 51, 52, 54, 96, 97, 98, 99, 100, 101, 102, 103, 105, 107, 108, 111, 113, 114, 116, 117, 118, 119, 120, 121, 122, 124, 125, 126, 127, 128, 129, 131, 230

F

Facebook, 75, 83
factor analysis, 26
Federal Communications Commission, 103, 228
Federal Government, 228
fertility, vii, 1, 54
fertilization, 3
fibroblasts, 50, 100, 119
filters, 231, 235, 236, 238, 240, 241, 242
financial, 58, 63, 67, 76, 181, 231, 232, 233
financial institutions, 232
financial performance, 181
financial records, 63
financial resources, 58, 67
Finland, 156, 162, 169, 171, 172, 174, 178
first generation, 88, 229
flexibility, 139, 140, 150
fluorescence, 21, 22, 29
folic acid, 124
follicles, 21, 22, 24, 26, 27, 28, 29, 30, 36, 51, 52
force, 37, 38, 39, 42, 43, 82
foreign firms, 169

formal education, ix, 133, 135, 136, 137
formaldehyde, 22, 28, 29, 129
formation, vii, 1, 44, 108, 118, 119, 120, 130, 139, 184
fragments, 97, 121
France, 56, 64, 156, 160, 161, 162, 169, 170, 171, 172, 173, 174, 178, 179, 235, 237
fraud, 236
free radicals, 44, 45
freedom, 57, 64, 81, 139, 141, 148, 185
functional changes, 98
funding, 238
funds, 229, 233
fusion, x, 124

G

gambling, x, 227, 228, 230, 231, 232, 233, 236, 237, 238, 239, 240, 241, 242
gametogenesis, 3, 21
gel, 97, 104, 118, 121, 123, 127, 129, 130
gene amplification, 108
gene expression, 99, 102
genes, 50, 52
genetic factors, 97
genetics, 122
genomic instability, 97, 100, 102, 124, 125
Georgia, 178, 179
germ line, 122
Germany, 22, 28, 29, 156, 157, 160, 161, 162, 163, 165, 169, 170, 171, 173, 174, 175, 178, 181
gestures, 82
gland, 100, 117
glioma, 101, 117, 126, 129
global communications, 57
global scale, 103
Global System for Mobile telecommunications (GSM), vii, 1
global village, 57, 58
globalization, 176, 177, 179
glucose, 186
glycerol, 22, 28
gonads, 22, 34, 36
governments, 233
GPS, x, 58, 79, 84
Greece, 1, 136, 156, 160, 161, 162, 172, 173, 175
group processes, 137
growth, 103, 155, 157, 158, 163, 164, 165, 176, 177, 179, 187, 188
GSM radiation exposure, vii, 2
guanine, 102, 125, 127
Guatemala, 177, 178
Guinea, 177, 178

H

handheld devices, 152
harmful effects, 100
hazards, 96, 97, 117, 120, 121, 130
health, vii, viii, ix, 2, 3, 34, 47, 48, 95, 96, 98, 99, 100, 101, 103, 107, 117, 118, 120, 121, 123, 126, 128, 130, 149, 183, 184, 185, 186, 187, 188, 189, 234
health care, 185, 187, 188, 234
health education, 188
health effects, 2, 100, 101, 107, 118, 120, 121
health information, 186
health problems, 184
health risks, 34, 117
health services, vii, ix, 183, 184
heart disease, 184
heat shock protein, 50, 52, 101, 118, 126
height, 4, 103, 106, 114, 115
heterogeneity, 140
high blood cholesterol, 185
high blood pressure, 184, 185
histone, 128
HIV-1, 127
House, 49, 50, 91
housing, 232
hTERT, 124
human, vii, 1, 2, 9, 30, 34, 35, 36, 41, 47, 48, 49, 50, 52, 54, 56, 62, 64, 67, 68, 71, 73, 74, 76, 89, 96, 97, 98, 99, 100, 101, 102, 104, 105, 114, 116, 117, 118, 119, 122, 123, 124, 125, 126, 127, 128, 129, 130, 131, 134, 138, 240
human body, 41, 89
human health, 48, 98, 99, 123, 130
human resources, 67, 71
humidity, 12
Hungary, 156, 160, 161, 162, 163, 164, 169, 170, 171, 172, 173, 174, 175
Hunter, 185, 188, 236, 240
hybrid, 153
hydrogen peroxide, 44
hydroxyl, 44
hypermedia, 65, 66, 67
hypertension, 107, 186
hypertext, 62, 65, 66
hypothesis, 43, 188

I

Iceland, 156
icon, 61, 78, 146
identification, 69, 96, 97, 98, 130, 234, 240, 241
identity, 138, 232, 233, 234
image, 57, 62, 63, 75, 105, 108, 134, 135, 139, 145, 146, 147, 172, 175, 241
image analysis, 105
image files, 105
imagination, 59, 62, 139
immune system, 51
improvements, 150, 151, 229
in vitro, 2, 36, 49, 50, 54, 97, 98, 100, 104, 116, 118, 119, 121, 123, 124, 129, 130, 131
in vitro exposure, 36, 54, 100, 116, 131
in vivo, 2, 35, 36, 54, 118, 120, 121, 124, 125, 129, 131
inappropriate content, x, 227, 228, 230, 231, 237, 238
incidence, viii, 95, 117, 127, 128
increased access, 228
incubation period, 49
India, 95, 96, 98, 103, 122, 125, 178, 179
individuals, viii, x, 71, 95, 97, 104, 121, 127, 129, 136, 137, 139, 141, 183, 184, 185, 227, 228, 230, 237, 238
induction, vii, 1, 2, 3, 21, 22, 23, 24, 25, 26, 28, 30, 34, 35, 36, 43, 44, 97, 99, 100, 101, 116, 118, 121, 127, 131
industrial environments, 96
industries, 232
industry, 48, 96, 97, 103, 155, 156, 157, 176, 181, 184, 185, 233, 236
inertia, 71
infertility, vii, 1, 2, 34, 35, 49, 121
information exchange, 187
infrastructure, x, 64
injury, 97, 128, 185
insects, 5, 7, 9, 12, 19, 21, 22, 23, 24, 26, 28, 29, 30, 31, 33, 34, 35, 36, 43, 45
insertion, 141
institutions, 136, 138, 156, 232
integration, 81, 158, 169
integrity, 49, 103, 108, 122
intensity values, 27, 106
interaction process, 69
interdependence, 66
interface, 67, 69, 76, 78, 80, 82, 84, 135, 140, 141, 142, 143, 144, 145, 146, 147, 148, 151, 152
interference, 18, 102
International Agency for Research on Cancer (IARC), viii, 95, 98
internationalization, vii, ix, 155, 156, 159, 160, 162, 163, 165, 169, 172, 176, 179, 180, 181
interphase, 97
interrelations, 74
intervention, 98, 105, 123

intervention strategies, 105
inventions, 57, 76
inventors, 58
investments, 67, 164, 169
ion channels, 18, 35, 36, 37, 38, 40, 42, 44
ionization, 102, 117
ionizing radiation, 3, 30, 43, 44, 96, 98, 99, 101, 103, 124, 126, 128
ions, 37, 38, 39, 40, 41, 42, 43, 47
IP address, 231, 232
Ireland, 156, 160, 161, 163, 165, 169, 170, 171, 173, 174, 175, 191
iron, 44
irradiation, 99, 104, 124
issues, vii, viii, 64, 69, 74, 95, 96, 123, 134, 136, 148, 149, 151, 231, 233, 235, 239, 242
Italy, 55, 57, 58, 63, 64, 69, 72, 73, 86, 151, 156, 161, 162, 170, 172, 173, 175
Ivory Coast, 177, 178, 179

J

Japan, 22, 29, 82, 152, 184
Java, 83, 144
Jordan, 99, 124, 153, 177, 178

K

K^+, 42, 52
Kazakhstan, 178, 179
Kenya, 177, 178
kinetics, 97
kinetochore, 122

L

labeling, 50
labor market, 70
landscape, 169
languages, 82
lasers, 29
Latin America, 177, 179
Latvia, 178, 179
laws, 70, 72, 187, 231, 232, 238
LEA, 92
lead, ix, 3, 36, 43, 44, 45, 47, 51, 56, 58, 61, 69, 78, 98, 99, 100, 102, 103, 119, 133, 134, 176, 230, 232
learners, 134, 137, 139
learning, vii, ix, 85, 133, 134, 135, 136, 137, 138, 139, 145, 147, 148, 149, 150, 151, 152, 153

learning environment, 135, 139, 149, 150, 152
learning process, 135, 138, 150
Lebanon, 177, 178
legal issues, 233
legality, 233
leisure time, 56, 57
lens, 52, 54, 101, 126
lesions, viii, 95, 99, 121, 130
leukemia, 99, 100, 119
liberty, 56
lifelong learning, 135, 151
lifetime, 138
light, 12, 77, 105, 125
lipids, 44
lithium, 61
Lithuania, 178, 179
lymphocytes, 49, 52, 100, 102, 104, 116, 117, 118, 124, 125, 127, 128, 129, 130, 131

M

Macedonia, 93, 178, 179
macromolecules, 44, 102
magnetic field, 9, 12, 13, 17, 19, 20, 37, 47, 49, 53, 99, 102, 117, 118, 125, 126, 127, 129
magnitude, 56, 71
major depression, 186, 189
majority, viii, 2, 24, 74, 82, 101, 156, 169, 171, 175, 181
Malaysia, 178, 179
mammalian cells, 27, 35, 128, 129
mammals, 35
man, viii, 95, 99
management, 67, 82, 145, 148, 183, 184, 185, 186, 187, 188, 189
manipulation, 58
manufacturing, 172
market share, 159, 176
market structure, 172
marketing, 64, 79, 80, 172, 230, 241
marrow, 104, 117, 125
mass, 30, 37, 45, 46, 103, 158, 189, 229
materials, 76, 79
matter, 20, 47, 103, 120, 139, 144, 228, 232, 235
measurement, x, 74, 128, 186
measurements, 5, 9, 12, 30, 126
media, 56, 57, 96, 139, 140, 145, 146, 147, 148, 149, 234, 240
medical, 104, 125, 185, 186, 188
medical care, 188
medical history, 104
medicine, ix, 183, 184, 188
Mediterranean, 58, 60, 62, 67, 73, 89

meiosis, 28
melanoma, 102
membranes, vii, 2, 3, 18, 35, 36, 39, 41, 43, 44
memory, 60, 79, 117
mental health, 187
mental model, 139
mercury, 4
mergers, 172
messages, x, 56, 58, 62, 83, 84, 87, 90, 139, 140, 141, 142, 143, 144, 145, 147, 148, 149, 227, 230, 239, 241
messenger RNA, 53
meter, 30, 107
methodology, viii, 13, 22, 28, 55, 62, 65, 67, 68, 71, 85, 89
methylation, 128
microbial cells, 100
micrometer, 105
micronucleated cells (MNC), viii, 95, 102
micronucleus (MN), 97, 98, 102, 105, 108, 119, 122, 123, 124, 125, 127, 128, 129, 130, 131
Micronucleus (MN) Assay, viii, 95
microscope, 22, 105
microscopy, 21
Microsoft, 74, 78, 80, 81, 82, 106
microwave radiation, vii, 1, 3, 19, 30, 36, 41, 44, 100, 116, 121, 124, 131
microwaves, 2, 35, 36, 43, 49, 50, 99, 100, 117, 122, 126, 128
migration, 103, 108
military, 58, 100
miniaturization, 60
mission, 48
missions, 5, 9
mitochondria, 44
mitosis, 105
mobile communication, viii, 95, 99, 101, 124, 156, 157, 176, 181, 184
mobile device, ix, x, 90, 133, 134, 135, 137, 138, 140, 145, 147, 148, 149, 150, 184, 187, 229, 230
Mobile health services, ix, 183
mobile telecommunication, 3, 36, 96, 181, 184
mobile telephony, vii, viii, 1, 2, 11, 12, 13, 14, 15, 17, 18, 20, 21, 27, 28, 30, 31, 32, 33, 34, 41, 43, 45, 47, 48, 95, 99, 101, 103, 127
models, vii, viii, 1, 3, 54, 55, 56, 58, 59, 60, 61, 66, 69, 74, 76, 78, 79, 80, 81, 84, 86, 87, 88, 99, 139
modern society, 121
modus operandi, 67, 73, 77
Moldavia, 178
Moldova, 178, 179

molecular biology, 101
molecules, 43, 44, 47, 117, 122
momentum, 122
money laundering, 236
monopoly, 62, 158
Montenegro, 178, 179
morphogenesis, 51
morphology, 29
mortality, 4, 35, 51, 124, 127, 189
mortality rate, 51
Moscow, 56
Moses, 235, 236, 241
motivation, 66
mRNAs, 29
mucosa, 104, 108, 109, 110, 118, 123, 128, 129
multimedia, viii, ix, 55, 56, 57, 59, 60, 62, 64, 65, 66, 67, 69, 72, 75, 76, 77, 78, 79, 80, 83, 84, 85, 86, 88, 89, 133, 135, 139, 141, 145, 147, 148, 149, 150, 151, 184
music, 88, 90, 91
mutations, 28, 98, 102, 101, 104, 127
MySpace, 241

N

national identity, 232
national security, 58
Nauru, 126
nausea, 107
necrosis, 28, 43
nerve, 18, 43, 45, 50
nervous system, 49, 100
Netherlands, 156, 160, 161, 162, 169, 170, 171, 172, 173, 174, 175, 181
neuroblastoma, 52
neuroma, 101, 127
neuroprotection, 124
neurotransmitter, 122
New Zealand, 178, 179
next generation, 28, 228
Nicaragua, 177, 178
non-ionizing electromagnetic radiations, viii, 95
North America, 77
Norway, 156, 162, 163, 164, 169, 171, 174, 175, 178
nucleation, 120
nuclei, 97, 105, 106, 108, 110
nucleic acid, 44
nucleus, 108, 110, 111
nutrients, 21
nutrition, 21, 27, 35

Index

O

obstacles, 20
Oceania, 180
octane, 22, 28
Olive tail movement, ix, 96
omission, 73
oocyte, 21, 24, 26, 27, 28, 29
oogenesis, 3, 4, 7, 21, 22, 24, 26, 27, 28, 30, 50, 52, 101
operating system, 75, 76, 78, 80, 81, 82, 83, 89
operations, 46, 68, 69, 70, 76, 82, 83, 84, 87, 90, 169
opportunism, 76
opportunities, ix, 133, 134, 137, 138, 148
organ, 118, 135
organelles, 43, 99, 129
organism, 21, 27, 28, 30
organize, 136, 140, 141, 144, 148
organs, 48
ornithine, 102
oscillation, 37
ovaries, 4, 21, 22, 26, 36
overlap, 68, 101
overproduction, 44
oviduct, 21
ownership, x, 156, 159, 169, 171, 181
oxidation, 102
oxidative damage, 124, 125
oxidative stress, vii, 1, 2, 3, 35, 36, 49, 52, 125, 128
oxygen, vii, 1, 36, 44, 50, 52, 54, 102, 118

P

Pakistan, 178, 179
paradigm shift, 126, 242
parallel, 5, 31, 41
parallelism, 59, 68
parental control, 236
Parliament, 239
participants, 139, 141, 142, 143, 144, 149, 186
password, 234
patient care, ix, 183
peripheral blood, 102, 117, 129, 131
permeability, 18, 50
personal communication, 56
personal computers, 59, 150
Peru, 177, 178
Philadelphia, 51
phosphate, 22
physical environment, x
physical health, 189
physical inactivity, 186
physicians, 186
physics, 96
Physiological, 21
pilot study, 49
plasma membrane, 43
platform, 105, 238
Plato, 91
Poland, 156, 160, 161, 162, 169, 170, 171, 173, 174
polar, 47
polymorphisms, 125
population, viii, ix, 2, 28, 36, 49, 56, 60, 62, 64, 70, 73, 77, 79, 81, 96, 99, 101, 103, 107, 111, 115, 116, 117, 118, 120, 123, 155, 157, 183, 184, 188, 189, 228
portability, 58
Portugal, 58, 156, 160, 161, 170, 172, 173
positive correlation, ix, 96, 103, 111, 113, 119, 120
potassium, 41, 49, 52
practical knowledge, 136
prestige, 172
prevention, 96, 130, 184, 185
principles, 62, 72, 74, 135
prisons, 237
probability, 6, 7, 10, 16, 19, 26, 31, 70, 72, 137
probe, 30
production costs, 62, 74
professionals, 81, 84, 148, 149, 186, 187, 188
profit, 63, 79, 83
progenitor cells, 52
programming, 64, 67, 75, 236
project, 58, 69, 124
proliferation, 99, 101, 102, 118, 128, 129
proposition, 119, 137
protection, 48, 98, 242
protein synthesis, 102
proteinase, 22, 29
proteins, 29, 37, 39, 44, 46, 50, 99, 102, 124
prototype, ix, 61, 133, 135, 140, 141, 144, 145, 148, 149, 150, 151, 152
prototypes, 62
psychology, 64
public health, 96, 103, 130
purchasing power, 56, 60, 88
P-value, 16
pyrimidine, 102

Q

quality standards, 77
quantification, 100, 127

R

race, 48, 228
radiation therapy, 96
radical formation, 44
radicals, 44, 45, 124
radio, viii, 49, 56, 57, 58, 59, 95, 96, 98, 99, 100, 103, 118, 129, 130, 155, 157, 158
radiography, 123
radiotherapy, 100, 127
radius, 31
radon, 101
random errors, 74
random sampling, 72
reactive oxygen species (ROS), vii, 1, 36, 44, 50, 52, 54, 102, 118
reactivity, 52
reading, x, 13, 30, 78, 81, 82
real time, 82, 228, 229
reality, 12, 59, 64, 66, 75, 126, 159, 234
reception, 5, 9, 11
receptors, 57
recovery process, 97
redistribution, 131
regeneration, 97
registries, 237
regulations, 130, 240
regulatory agencies, 103
rejection, 74, 173, 189
reliability, x, 62, 84, 97, 236
remodelling, 76
repair, 28, 51, 97, 99, 100, 102, 104, 106, 120, 121, 122, 123, 124, 125, 126, 127, 129, 130
representativeness, 70, 71, 72, 73
reproduction, vii, 1, 5, 6, 21, 33, 34, 44, 58, 80, 84
reproductive cells, vii, 1, 2, 27, 28, 34, 35, 36, 44
reproductive organs, 48
researchers, ix, 58, 69, 72, 133, 134
residues, 52, 102
resolution, 58, 68, 69, 144
resonator, 100
resources, ix, 58, 66, 67, 68, 69, 71, 78, 133, 135, 139, 140, 144, 145, 147, 148, 149, 150, 151, 183, 184
response, 28, 98, 100, 124, 125, 127, 129, 231, 232, 233
restoration, 37
restrictions, 81, 120, 135, 150, 156, 158, 233, 236
reticulum, 43
revenue, 230, 232, 238
risk, viii, 51, 95, 96, 98, 99, 100, 101, 117, 121, 122, 124, 126, 127, 129, 130, 185, 186, 233
risk assessment, 51
risk factors, 121, 185
risks, 34, 96, 99, 117
RNAs, 53, 118
Romania, 178, 179
rules, 77, 187, 242

S

safety, 18, 34, 48, 81, 84, 120, 188, 238
salivary gland, 100, 117, 122
sample design, 70
sampling error, 71, 73
Samsung, 60, 77, 79, 88
satellite technology, 228
scale economies, 158
school, 48, 88, 134, 135, 137, 138, 149, 188, 230
scientific method, 68
scientific progress, 74
scope, 66, 136, 137, 156, 159, 160, 163, 172, 177, 180
Second World, 58
security, 58, 234, 237, 240, 242
security forces, 237
self-awareness, 241
self-organization, 144
self-verification, 231
semen, 49, 121
semiconductor, 61
semiotics, 65, 66
sensations, 100
senses, 89
sensing, 52
sensitivity, 104, 117
sensors, 37, 38, 42, 89
Serbia, 178, 179
servers, 232
service provider, 155, 186, 231, 232
service quality, 155
services, 63, 64, 75, 77, 79, 81, 83, 134, 137, 153, 158, 162, 177, 183, 184, 185, 186, 188, 229, 230, 235, 236, 238
sex, viii, 7, 8, 86, 95, 104, 106, 111, 119, 236
shape, 45, 58, 61, 89, 114
shelf life, 83
shock, 21, 27, 50, 52, 101, 118, 126
showing, ix, 100, 108, 109, 110, 114, 133, 145, 147
siblings, 88
signal transduction, 102
signals, x, 5, 9, 11, 12, 15, 17, 18, 30, 31, 35, 36, 41, 42, 43, 48, 54, 58, 101, 117, 118
signs, 99, 241
silicon, 240
simulations, 46, 47

sister chromatid exchange, 97, 118
sleep disturbance, 2, 35
sleeping problems, 51
Slovakia, 156, 178, 179
smoking, 104, 106, 111, 118, 119, 128, 129, 185
SMS, 56, 60, 62, 64, 65, 87, 88, 92, 229, 236
sociability, 189
social context, 68
social group, 57
social network, vii, 58, 64, 134
social participation, 138
social psychology, 64
social sciences, 66, 67, 68, 69
social support, 184, 186, 187, 188
social support network, 184
social theory, 138
socialization, 56, 57, 134
society, 59, 96, 99, 121, 134, 157, 184, 189, 234, 239
software, ix, 56, 57, 62, 65, 66, 74, 75, 76, 77, 78, 79, 80, 81, 82, 83, 89, 105, 106, 133, 135, 137, 138, 141, 144, 145, 148, 149, 150, 229, 231, 232, 236, 237, 240
solution, 22, 28, 29, 37, 79, 84, 100, 135, 136, 137, 140, 141, 143, 144, 149
somatic cell, 36
South America, 175
South Asia, 180
Spain, 52, 56, 58, 59, 63, 64, 69, 76, 86, 155, 156, 160, 161, 162, 163, 164, 169, 170, 171, 173, 174, 175, 178, 179
spam, 240
specialists, 72
species, vii, 1, 36, 44, 50, 52, 54, 102, 118
Specific Absorption Rate (SAR), vii, 2, 9, 30, 103
specific heat, 46
speech, 143, 144, 235
sperm, 36, 52, 54
spermatogenesis, 3, 4, 7, 21
Sprague-Dawley rats, 101, 125
stakeholders, 101, 235
standard deviation, 9, 10, 12, 15, 17, 26, 31, 33, 73, 110, 114
standard error, 110, 114
standardization, 155, 156
state, 38, 39, 62, 80, 104, 118, 141, 144, 232
state enterprises, 62
states, 40, 96, 137, 140, 143, 241
statistics, viii, 55, 57, 67, 70, 71, 72, 103
stimulus, 30
storage, 140
stress, 1, 2, 3, 21, 25, 27, 28, 35, 36, 49, 52, 88, 99, 117, 125, 128
stress factors, 21, 27, 35

structure, 64, 67, 72, 97, 99, 172, 236
style, 78, 80, 121, 230, 241
subscribers, 64, 103, 156, 157, 158, 163, 164, 165, 176, 236
success rate, 236
supervision, 237
Supreme Court, 235, 241
surveillance, 237
survival, 131
Sweden, 51, 156, 160, 161, 162, 163, 164, 169, 170, 171, 172, 173, 174, 175, 178, 181
Switzerland, 156, 161, 162, 169, 171, 172, 174, 175
symptoms, vii, 1, 3, 51, 100, 125
synchronization, 82
syndrome, 35, 129, 131
synergistic effect, 117, 129
synthesis, 99, 102, 123, 127, 141

T

Taiwan, 79, 152, 182, 183
Tajikistan, 178, 179
talent, 68
target, 68, 70, 149, 160, 232, 236, 237
technician, 124
techniques, viii, 55, 57, 65, 66, 67, 68, 69, 71, 72, 88, 99, 101, 135
technological advances, 60
technologies, 56, 57, 60, 62, 64, 67, 76, 79, 83, 96, 98, 101, 120, 137, 151, 184, 228, 238
technology, iv, vii, ix, x, 48, 56, 61, 76, 78, 79, 81, 96, 134, 139, 149, 155, 157, 158, 159, 160, 183, 184, 188, 227, 228, 229, 230, 235, 236, 237, 238
telecommunications, vii, 1, 3, 36, 57, 62, 64, 75, 76, 79, 90, 96, 159, 181, 232
telephones, viii, 48, 52, 54, 95, 96, 99, 100, 101, 122, 124, 125, 126, 127, 130, 131, 157
temperature, vii, 2, 4, 12, 15, 32, 33, 35, 41, 46, 106, 186
terminals, 58, 239
territory, 57
testing, 66, 87, 98, 129, 240
text messaging, 184, 188
text mining, 236, 239
Thailand, 178, 179
therapy, 96, 98
thermal energy, 47
threats, 242
thymus, 100, 128
time constraints, 134
tissue, viii, 2, 30, 37, 45, 46, 47, 49, 51, 97, 100, 102, 103, 125, 130

tobacco, 129
topology, 76
total micronuclei (TMN), viii, 95
tourism, 62, 65, 85
toxicology, 97, 124, 129
toxicology studies, 124
trade, 81, 232
training, 56, 70, 72, 136
trajectory, 75
transactions, 231, 232, 233
transcription, 118
transduction, 102
transformation, 98, 119, 128
translation, 69
transmission, x, 5, 12, 46, 100, 107, 186, 227, 228, 229
transport, 29
transportation, 86
treatment, viii, ix, 5, 31, 95, 183, 184, 185
tumor cells, 125
tumors, 36, 51, 117, 122, 124, 125, 128
tumour treatment, viii, 95
tumours, 51, 101, 117, 125
Turkey, 178, 179

U

Ukraine, 178, 179
United Kingdom (UK), 122, 129, 152, 156, 157, 160, 161, 162, 163, 165, 169, 170, 171, 173, 174, 175, 178, 179, 181, 235, 236, 240
United Nations, 184, 188, 189
United States (USA), 9, 22, 28, 49, 103, 122, 153, 178, 179, 181, 228, 230, 232, 233, 235, 239, 242
universe, 67, 68, 69, 70, 71, 72, 73, 86
university education, 73
updating, 56, 75, 78, 235, 236
uplink, 11
urban, x, 96
Uruguay, 58, 177, 178
UV, 99, 125
Uzbekistan, 178, 179

V

valence, 37, 39, 40, 42
validation, 97, 123
Vanuatu, 177, 178
variables, 16, 68, 69, 179
variations, 5, 6, 7, 10, 16, 19, 26, 31, 32, 118, 230, 234
VAT, 63
velocity, 37, 41
Venezuela, 177, 178, 179
venue, 231
vibration, 37, 38, 40, 41, 42, 47
videos, ix, 134, 135, 145
violence, 236
vision, 76
visualization, 66, 105
vitamin A, 129
vote, 140, 141, 143, 144

W

war, 124
Washington, 241
waste, 21, 68, 69
watches, 89
water, 39, 44
wave power, 46
wealth, 66
web, x, 57, 62, 187, 227, 236, 238, 239, 240
websites, 236, 237
weight loss, 187, 188
weight management, 187, 188
Wi-Fi, 185
WiFi signal strength, x, 191, 192, 195, 197
windows, vii, 2, 17, 18, 19, 43, 50, 78, 82
wireless networks, 180
wireless technology, 155, 158
workers, 101, 114, 118, 131
workplace, 236
World Bank, 182
World Health Organization (WHO), 96, 101, 103, 118, 120, 130, 131, 186, 189
worldwide, viii, 95, 96, 183, 184, 228

X

X-irradiation, 124
x-rays, 124

Y

young people, 184